# Lecture Notes in Mathematics

Edited by A. Dold and B. Eckmann

760



H. O. Georgii




# Canonical Gibbs Measures

## Some Extensions of de Finetti's Representation Theorem for Interacting Particle Systems


Springer-Verlag
Berlin Heidelberg New York 1979

**Author**

H. O. Georgii
Fakultät für Mathematik
Universität Bielefeld
Postfach 8640
D-4800 Bielefeld

AMS Subject Classifications (1970): 60K35, 82A60

ISBN 3-540-09712-0 Springer-Verlag Berlin Heidelberg New York
ISBN 0-387-09712-0 Springer-Verlag New York Heidelberg Berlin

Library of Congress Cataloging in Publication Data
Georgii, Hans-Otto.
Canonical Gibbs measures.
(Lecture notes in mathematics; v. 760)
Bibliography: p.
Includes index.
1. Probabilities. 2. Representations of groups. 3. Measure theory. I. Title. II. Title: Gibbs measures. III. Title: De Finetti's representation theorem. IV. Series: Lecture notes in mathematics (Berlin); v. 760.
QA3.L28 no. 760 [QC20.7.P7] 510'.8s [519.2] 79-23184
ISBN 0-387-09712-0



Printing and binding: Beltz Offsetdruck, Hemsbach/Bergstr.
2141/3140-543210

Contents

# Introduction

Symmetric probability measures on the infinite product

$$\Omega = F \times F \times \ldots$$

of a set $F$ (equiped with a $\sigma$ - algebra) are a well-known concept in probability theory: A probability measure on $\Omega$ is said to be symmetric if it is invariant under the group of those transformations of $\Omega$ which are defined by a permutation of finitely many coordinates. Apparently the first place in which this notion appeared was a contribution of J. Haag to the International Congress of Mathematicians at Toronto in 1924. B. de Finetti (1931) independently introduced the idea of symmetric probability measures. For the case $F = \{0 , 1\}$ he proved the following theorem which is now well-known: Each symmetric probability measure $\mu$ is a mixture of homogeneous product measures, i. e., $\mu$ has a representation

$$(0.1) \qquad \mu = \int m(d\alpha) \quad \alpha \otimes \alpha \otimes \ldots$$

where $m$ is a probability measure on the set of all probability measures on $F$. Moreover, $m$ is uniquely determined and is obtained as the limit distribution of the empirical distributions. (It is worthwile mentioning that de Finetti was interested in this result for philosophical reasons: It shows that a statement of the form "I believe that the tosses of this coin are independent and identically distributed, with the unknown probability of heads occuring lying somewhere between 1/3 and 2/3" is equivalent to the purely subjective statement "I believe the tosses are symmetrically distributed and that the frequency of heads will fall somewhere between 1/3 and 2/3".)

De Finetti's theorem was extended to more general $F$'s by A. Khintchine (1932, 1952), B. de Finetti (1937), E.B. Dynkin (1953) and finally E. Hewitt and L.J. Savage (1955) (this last paper should be consulted for the earlier references).

Hewitt and Savage pointed out that the representation (0.2) results from a combination of the following two assertions: Each symmetric $\mu$ is the mixture of extreme symmetric probability measures, and the extreme symmetric probability measures are just the homogeneous product measures. They proved that the second assertion is true without any condition on $F$, and found general conditions implying the first. (Their famous 0 - 1 law is equivalent to the statement that the homogeneous product measures are extreme symmetric measures; L.E. Dubins and D.A. Freedman (1979) have shown that the first assertion may fail to hold even when $F$ is a separable metric space.) There are at least two lines of further research which arose from their paper. The first line was concerned with the question of whether the 0 - 1 law could be extended to products of not necessarily identical probability measures, see, for instance, D. Aldous and J. Pitman (1977), G. Simons (1978) and the references therein. A second group of papers dealt with the problem of whether de Finetti's representation theorem has an extension to probability measures with a weaker symmetry condition; for instance a condition which is satisfied by all mixtures of certain Markov chains; see D.A. Freedman (1962), T. Höglund (1974), S.L. Lauritzen (1974), and P. Martin-Löf (1974).

Recently, and independently of this statistical tradition, the need for such representation theorems also arose in Statistical Mechanics. It is the purpose of this text to explain the origins of this need and to give some of the required theorems. So let us describe the problem. Let $F = \{0, 1\}$, choose a countably infinite set $S$ and regard the product $\Omega = F^S$ as the space of all configurations of indistinguishable particles in $S$, no two of which are allowed to occupy the same site. If (apart from this exclusion rule) the particles do not interact but possibly prefer certain sites then in equilibrium the state of this particle system would be described by a (not necessarily homogeneous) product measure. Clearly, from a physical point of view it is much more natural to consider particle systems with an interaction. Then, as well as specifying a self-potential which describes to what extent the particles prefer to stay at each site, it is also necessary to give the additional energy required in order that a pair of sites should be occupied. (This, of course, is assuming only a pairwise interaction.) In this case the set of equilibrium states (which is possibly not a singleton!) is given by the set of all so-called (grand canonical) Gibbs measures for this interaction. These are defined as those probability measures which have prescribed versions (depending on the interaction) of their conditional probabilities with respect to the configurations outside each finite region.

In certain situations, however, it is only the interaction potential which is determined by the physical circumstances and not the self-potential. For instance, suppose we have a time evolution of the particle system in which the particles may change their positions but cannot be created or destroyed. Moreover, assume that the particle motion is governed by the interaction in the sense that those particle jumps are favoured which entail the largest gain of total energy. In order to establish that such an evolution is locally in equilibrium it is sufficient to know that in each finite region the configurations have a particular distribution (given in terms of the interaction) *when in addition to the configuration outside the region the particle number in the region is also fixed*. We call a probability measure with this property a *canonical Gibbs measure* because these specific local equilibrium distributions corresponding to given environments and particle numbers are just the so-called canonical Gibbs distributions. If there is no interaction the canonical Gibbs measures are exactly the symmetric measures.

In this text we will ask whether the analogue of de Finetti's result holds, namely whether each canonical Gibbs measure is a mixture of measures for which the distributions of the local particle numbers also have a particular form (being defined in terms of a self-potential), i. e., a mixture of grand canonical Gibbs measures. We will show that the answer is in the negative when the interaction and the self-potential are spatially very inhomogeneous but is positive as soon as these are, in some sense, sufficiently homogeneous. In our framework there will be no difficulty to show that each canonical Gibbs measure is a mixture of extreme canonical Gibbs measures. Thus our main task will be to find natural conditions which ensure that each extreme canonical Gibbs measure is a (grand canonical) Gibbs measure. As a particular result we will obtain some extensions of the Hewitt/Savage 0 - 1 law to inhomogeneous non-product measures.

Thus far we have only described particle systems on a discrete set. But clearly the same questions arise if the position space of the particles is continuous, and of course this is the case of most interest in physics. Therefore we will also be concerned with canonical and grand canonical Gibbs point processes, and we will obtain some extensions of the result stated immediately below which is the point process counterpart of de Finetti's theorem, and was first proved by K. Nawrotzki (1962) and D.A. Freedman (1963) . Suppose $\mu$ is a point process on the real line (i. e., a probability measure on the set of all locally finite point configurations on $\mathbb{R}$ ) satisfying the following symmetry condition: For each bounded interval $\Lambda$ and a fixed number of particles in $\Lambda$ the positions of these particles are independently and uniformly distributed. Then $\mu$ is a mixed Poisson process, i. e., $\mu$ has a representation

(0.2) $$\mu = \int m(dz) \; \pi^z \quad ,$$

where $m$ is a (uniquely determined) probability measure on $[0, \infty[$ and $\pi^z$ denotes the Poisson point process on $\mathbb{R}$ with intensity $z \geq 0$ .

The notion of a canonical Gibbs measure has an obvious generalization, namely the concept of a microcanonical Gibbs measure. To obtain its definition we have only to change the italicized phrase above "when ... the particle number in the region is also fixed" into "when ... the values of certain extensive quantities (as, for example, the particle number and the interaction energy) in the region are also specified" . However, we think it is reasonable to confine ourselves to canonical Gibbs measures. The reader interested in the microcanonical case is referred to M. Aizenman et al. (1978) , C. Preston (1978) , and R.L. Thompson (1974) .

I am much indepted to C. Preston for help with the English and to U. Sander for typing the manuscript.

## § 1 Basic concepts

We will consider both discrete and continuous models for an interacting particle system. In both cases we restrict ourselves to the situation which is usually considered in classical statistical mechanics. In the discrete models the particles are allowed to occupy the sites of some countably infinite set $S$, e.g. the three-dimensional integer lattice. If the particles are indistinguishable, then at each site $x \in S$ we have the alternatives: $x$ is occupied by a particle or not. To stress the symmetry between these two possibilities we imagine the empty sites to be occupied by particles of a species $0$. More generally, we will consider an arbitrary finite set $F$ of particle types. Then the space of all particle configurations is $\Omega = F^S$.

In the continuous case we assume that there is only one type of particle and the particles take their positions in a nice subset $S$ of a Euclidean space. If we exclude the possibility that two particles have the same position the set $\Omega$ of all locally finite subsets of $S$ becomes the configuration space. (The generalization to more general spaces $S$ will be clear whenever it is possible. Furthermore, the case of several types of particles can be reduced to the case of a single type by embedding several copies of $S$ in a higher dimensional space.)

In order to completely formulate the models, in both the discrete and the continuous case, it is necessary to specify the interaction between the particles. Then (under a hypothesis of thermodynamic equilibrium) the commonly accepted Ansatz of Gibbs yields the probability distributions for local configurations conditioned with respect to a fixed environment. The objects of our study are probability measures on $\Omega$ whose local behaviour is determined by the Gibbs distributions.

We should mention that a unified treatment of both the discrete and the continuous case would be possible using the formalism of point process theory or the abstract setting of Preston (1976, 1978). However, we think that each case is better treated separately, and this facilitates the possible study of only one of the cases.

## 1.1 The discrete model

Let us start with $S$ , a countably infinite set of particle sites, and $F$ , a finite set of types of particles with cardinality $|F| \geq 2$ . The configuration space is $\Omega = F^S$ .

For each $V \subset S$ let

$$(1.1) \qquad X_V : \omega = (\omega_x)_{x\in S} \rightarrow \omega_V = (\omega_x)_{x\in V}$$

be the projection from $\Omega$ on $\Omega_V = F^V$ . We use the same symbol for the projection from $\Omega_W$ to $\Omega_V$ whenever $W \supset V$ . If $V$ and $W$ are disjoint subsets of $S$ and $\omega \in \Omega_V$ , $\zeta \in \Omega_W$ , we denote by $\omega\zeta$ the configuration on $V \cup W$ with $(\omega\zeta)_V = \omega$ and $(\omega\zeta)_W = \zeta$ .

For any $a \in F$ , $\omega \in \Omega_V$ let

$$(1.2) \qquad N(a, \omega) = |\{x \in V: \omega_x = a\}|$$

be the number of $a$ - particles in the configuration $\omega$ . (Which $V$ has to be used will be clear from the context. If $\omega \in \Omega$ and only the particles in $V$ are counted we write $N(a, \omega_V)$ . ) Finally, let

$$(1.3) \qquad N(\omega) = (N(a, \omega))_{a\in F} .$$

We let

$$(1.4) \qquad \mathcal{S} = \{\Lambda \subset S : 0 < |\Lambda| < \infty\}$$

denote the system of all non-empty finite subsets of $S$ . For singletons in $\mathcal{S}$ we usually write $x$ instead of $\{x\}$ . Often we use the symbol

$$\lim_{\Lambda \uparrow S} \tag{1.5}$$

with the meaning that the limit is taken over a fixed sequence $(\Lambda_n)_{n\geq 1}$ in $S$ such that $\Lambda_n \subset \Lambda_{n+1}$ $(n \geq 1)$ and $\bigcup_{n\geq 1} \Lambda_n = S$. This sequence may be chosen arbitrarily unless the contrary is stated.

For any $\Lambda \in S$ let

$$A_\Lambda = \{L \in \{0, 1, 2, \ldots\}^F : \sum_{a\in F} L(a) = |\Lambda|\} \tag{1.6}$$

be the range of the function $\omega \rightarrow N(\omega_\Lambda)$, and for $L \in A_\Lambda$ let

$$\Omega_{\Lambda,L} = \{\omega \in \Omega_\Lambda : N(\omega) = L\} \tag{1.7}$$

be the set of configurations in $\Lambda$ with given $L$.

For any $V \subset S$ we denote by

$$F_V = \sigma(X_x : x \in V) \tag{1.8}$$

the $\sigma$ - algebra of the events in $V$ which are generated by the finite-dimensional projections in $V$. It is well-known that $F = F_S$ is generated by the product topology on $\Omega$ (with the discrete topology on $F$). Since $\Omega$ is compact the set of probability measures on $(\Omega, F)$ also is compact in the weak topology. Note that a sequence $(\mu_n)$ of probability measures on $(\Omega, F)$ converges weakly to some $\mu$ as soon as all cylinder probabilities converge:

$$\mu_n(X_\Lambda = \zeta) \rightarrow \mu(X_\Lambda = \zeta) \quad (\zeta \in \Omega_\Lambda, \Lambda \in S) .$$

The following sub - $\sigma$ - algebras of $F$ will play a central rôle: the tail field

(1.9) $$F_\infty = \bigcap_{\Lambda \in S} F_{S \setminus \Lambda}$$

and the $\sigma$ - algebra of symmetric events

(1.10) $$E_\infty = \bigcap_{\Lambda \in S} E_\Lambda \quad .$$

Here for any $\Lambda \in S$

(1.11) $$E_\Lambda = \sigma\,(\,N(X_\Lambda),\, F_{S \setminus \Lambda})$$

denotes the $\sigma$ - algebra of events which are invariant under permutations of the sites in $\Lambda$ . Note that $E_\Lambda \supset E_{\Lambda'}$ whenever $\Lambda \subset \Lambda'$ .

The objects which we are interested in are probability measures on $(\Omega, F)$ whose conditional probabilities with respect to either $F_{S \setminus \Lambda}$ or $E_\Lambda$ have a particular version. This version has the form postulated by Gibbs and is determined by a potential $\Phi$ describing the interaction of the particles.

(1.12) Definition: A function

$$\Phi : S \times \Omega \to \mathbb{R}$$

is called a *potential* if

(i) $\Phi(A, \cdot)$ is $F_A$ - measurable for any $A \in S$ . (Sometimes therefore we shall think of $\Phi(A, \cdot)$ as a function on $\Omega_A$ .)

(ii) for any $\Lambda \in S$ , $\zeta \in \Omega_\Lambda$ , $\omega \in \Omega_\Lambda$ the *energy* of $\zeta$ in $\Lambda$ with boundary condition $\omega$

$$E_\Lambda(\zeta \mid \omega) = \sum_{A \in S : A \cap \Lambda \neq \emptyset} \Phi(A, \zeta\, \omega_{S \setminus \Lambda})$$

is well-defined (as the finite limit of the partial sums over all $A \subset V$ if

$V$ runs through the directed set $S$ ) and continuous as a function of $\omega$ .

This continuous dependence of the energy on the boundary condition says that the potential decays sufficiently rapidly for sets $A$ with "large diameter" . A well-known sufficient condition for (ii) is

$$(1.13) \qquad \sum_{x \in A \in S} \| \Phi (A, \cdot ) \| < \infty \qquad (x \in S) \quad ,$$

where $\| \cdot \|$ denotes the sup-norm.

Without loss of generality we could assume that the potential vanishes for certain configurations, see e.g. Sullivan (1973). We shall not use such a normalization unless $F = \{0, 1\}$ . In this case we can assume that there is a function $S \to \mathbb{R}$ (again denoted by $\Phi$ ) such that

$$(1.14) \qquad \Phi (A, \omega) = \Phi (A) \, \omega^A \qquad (A \in S, \omega \in \Omega) \quad ,$$

where $\omega^A = \prod_{x \in A} \omega_x = 1_{\{X_A = 1\}} (\omega)$ .

The so-called chemical potential $\Phi (x, a)$ $(x \in S, a \in F)$ specifies to which extent the site $x$ favours the occurence of an $a$ - particle. In order to favour type - $a$ particles on the whole of $S$ against all other types it is sufficient to add a constant (written in the form $- \log z(a)$) to $\Phi( \cdot , a )$ . Since this procedure will be important for us we shall include it from the beginning:

(1.15) Definition: A function

$$z : F \to [0, \infty[ \quad \text{is called an } \textit{activity} \text{ if } \quad \sum_{a \in F} z(a) > 0 \quad .$$

$A$ denotes the set of all activities.

As we shall see at once in (1.16) , it makes no difference for us if some $z \in A$ is multiplied by a positive factor. Therefore we write $z_1 \sim z_2$ if $z_1 = c \, z_2$ for some $c > 0$ . In the case $F = \{0, 1\}$ the equivalence class of any $z \in A$ usually is represented by the number $z' \in [0, \infty]$ given by

$$z' = \begin{cases} z(1) \ / \ z(0) & \text{if } z(0) > 0 \\ \infty & \text{otherwise} \end{cases} .$$

In the following we fix a potential $\Phi$. Together with an activity $z \in A$, $\Phi$ defines a system of conditional probabilities as follows:

(1.16) Definition: Let $\Lambda \in S$, $\omega \in \Omega$. The probability distribution on $\Omega_\Lambda$ defined by

$$\gamma_\Lambda^z (\zeta|\omega) = Z_\Lambda (z, \omega)^{-1} \prod_{a\in F} z(a)^{N(a,\zeta)} \exp [-E_\Lambda (\zeta|\omega)] \qquad (\zeta \in \Omega_\Lambda)$$

is called the (*grand canonical*) *Gibbs distribution* on $\Lambda$ with boundary condition $\omega$ corresponding to the potential $\Phi$ and activity $z$. The normalization factor

$$Z_\Lambda (z, \omega) = \sum_{\zeta\in\Omega_\Lambda} \prod_{a\in F} z(a)^{N(a,\zeta)} \exp [-E_\Lambda (\zeta|\omega)]$$

is called the *partition function*.

It follows directly from the definition that the family of Gibbs distributions is consistent in the following sense:

(1.17) $$\gamma_\Delta^z (\zeta|\omega) = \gamma_\Lambda^z (\zeta_\Lambda|\zeta\omega_{S\setminus\Delta}) \ \gamma_\Delta^z (\{X_{\Delta\setminus\Lambda} = \zeta_{\Delta\setminus\Lambda}\}|\omega)$$

whenever $\Lambda \subset \Delta$, $\zeta \in \Omega_\Delta$, $\omega \in \Omega$. Conversely, any reasonable consistent system of conditional probabilities can be described in terms of a potential. The Möbius inversion formula shows (see, for example, Sullivan (1973)) :

(1.18) Remark: *Suppose that a system* $(g_\Lambda(\cdot|\omega))_{\Lambda\in S,\omega\in\Omega}$ *of conditional probabi-*

*lities is consistent in the sense of* (1.17) *and that the function* $g_\Lambda(\zeta|\cdot)$ *are* $F_{S\setminus\Lambda}$ *- measurable, continuous, and strictly positive. Then there is a potential* $\Phi$ *such that* $g_\Lambda = \gamma^1_\Lambda$ *for all* $\Lambda \in S$ .

If $\mu$ is a probability measure on $(\Omega, F)$ and $\Lambda \in S$ , $\zeta \in \Omega_\Lambda$ then we let

(1.19) $$\mu_\Lambda(\zeta|\cdot) = \mu(X_\Lambda = \zeta | F_{S\setminus\Lambda})$$

denote the conditional probability of the event $\{X_\Lambda = \zeta\}$ with respect to the $\sigma$ - algebra $F_{S\setminus\Lambda}$ and the measure $\mu$ .

(1.20) Definition: A probability measure $\mu$ on $(\Omega, F)$ is called a *Gibbs measure* with respect to the activity $z \in A$ and the potential $\Phi$ if for any $\Lambda \in S$ and $\zeta \in \Omega_\Lambda$

$$\mu_\Lambda(\zeta|\cdot) = \gamma^z_\Lambda(\zeta|\cdot) \qquad \mu\text{ - a.s.}$$

We let $G(z) = G(z,\Phi)$ denote the set of all such Gibbs measures.

This notion is due to Dobrushin (1968) and Lanford and Ruelle (1969). For this reason, Gibbs measures are often called DLR - states. They have given rise to an extensive theory describing the properties of many-particle systems in thermodynamic equilibrium. Certain aspects of this theory can be found in Preston (1973, 1976) and Ruelle (1978), for instance. In particular, one knows:

(1.21) Theorem: $G(z)$ *is always non-empty, convex, and weakly compact. For some* $\Phi$ *and* $z$ *we have* $|G(z)| > 1$ .

The non-uniqueness phenomenon $|G(z)| > 1$ has the physical interpretation of a phase transition and is therefore of particular interest. However, since this phenomenon is not the theme of this text we refer the reader to Dobrushin (1968 b), e.g.. Here we will identify the set $G(z)$ only in two simple situations:

(1.22) Example: *Suppose that* $\Phi(A,\cdot) = 0$ *whenever* $|A| > 1$ . *Then* $G(z)$

*is a singleton consisting of the product measure* $\pi^z$ *whose marginal distributions are given by*

$$\pi^z(X_x = a) = z(a) \exp[-\Phi(x, a)] / \sum_{b \in F} z(b) \exp[-\Phi(x, b)] \quad .$$

Indeed, it is easy to see that $\gamma_\Lambda^z(\cdot|\omega)$ is a product measure with these marginal distributions and does not depend on $\omega$ . Also, it is simple to verify:

(1.23) Example: *Let* $a \in F$ *and suppose that* $z(b) = 0$ *unless* $b = a$ . *Then for each* $\Phi$ , $G(z)$ *consists of the point mass* $\varepsilon_a$ *on the constant configuration whose value at each site is* $a$ .

What is the difference in the local behaviour of two Gibbs measures with respect to the same potential but different activities? Obviously, their behaviour will be different when an $a$ - particle at some site is replaced by a particle of a type $b \neq a$ . For some models describing the time evolution of a many particle system (the so-called birth - and - death or spin - flip processes, see Liggett (1977), for instance) only this kind of replacements occur during the evolution. Therefore, if a Gibbs measure is invariant under such an evolution then its activity is uniquely determined. A different type of process has been introduced by Spitzer (1970). It models a system of infinitely many moving particles, see section 2.1 for details. In this kind of time evolution, particles of different types interchange their positions, and the behaviour of a Gibbs measure under such interchanges depends only on its potential and not on its activity. Indeed, such interchanges do not alter the total particle numbers in sufficiently large regions, and for each $\Lambda \in S$ and $L \in A_\Lambda$ the conditional probability given $\Omega_{\Lambda,L}$ with respect to a Gibbs distribution $\gamma_\Lambda^z(\cdot|\omega)$ (if it is well-defined) does not depend on $z$ , and is given by $\gamma_{\Lambda,L}(\cdot|\omega)$ , which is defined as follows:

(1.24) Definition: Let $\Lambda \in S$ , $\omega \in \Omega$ , $L \in A_\Lambda$ . The probability distribution on $\Omega_\Lambda$ defined by

$$\gamma_{\Lambda,L}(\zeta|\omega) = 1_{\Omega_{\Lambda,L}}(\zeta)\, Z_{\Lambda,L}(\omega)^{-1} \exp[-E_\Lambda(\zeta|\omega)] \qquad (\zeta \in \Omega_\Lambda)$$

is called the *canonical Gibbs distribution* for $\Phi$ on $\Lambda$ with particle numbers $L(a)$, $a \in F$, and boundary condition $\omega$. The normalization factor

$$Z_{\Lambda,L}(\omega) = \sum_{\zeta\in\Omega_{\Lambda,L}} \exp[-E_\Lambda(\zeta|\omega)]$$

is called the *canonical partition function*. If $L \notin A_\Lambda$ we put $Z_{\Lambda,L}(\omega) = 0$.

The following consistency property, which is easily verified, makes precise the statement preceding the definition (1.24):

(1.25) $$\gamma_\Lambda^z(\zeta|\omega) = \gamma_{\Lambda,L}(\zeta|\omega)\ \gamma_\Lambda^z(\Omega_{\Lambda,L}|\omega)$$

whenever $z \in A$, $\Lambda \in S$, $L \in A_\Lambda$, $\zeta \in \Omega_{\Lambda,L}$, and $\omega \in \Omega$. Furthermore, the family of canonical Gibbs distributions satisfies the consistency condition

(1.26) $$\gamma_{\Delta,L}(\zeta|\omega) = \gamma_{\Lambda,N(\zeta_\Lambda)}(\zeta_\Lambda|\zeta\omega_{S\setminus\Delta})\ \gamma_{\Delta,L}(\{N(X_\Lambda) = N(\zeta_\Lambda), X_{\Delta\setminus\Lambda} = \zeta_{\Delta\setminus\Lambda}\}|\omega)$$

where $\Lambda \subset \Delta$, $L \in A_\Delta$, $\zeta \in \Omega_{\Delta,L}$, and $\omega \in \Omega$.

Similarly as in (1.20), the system of canonical Gibbs distributions can be used for the definition of canonical Gibbs measures. These will be the central notion of this text. In particular, we will see in § 2 that these measures are invariant under the particle motions mentioned above. Note that for any $\Lambda \in S$ and $\zeta \in \Omega_\Lambda$ the function

$$\omega \to \gamma_{\Lambda,N(\omega_\Lambda)}(\zeta|\omega)$$

is measurable with respect to $E_\Lambda$.

(1.27) <u>Definition:</u> A probability measure $\mu$ on $(\Omega, F)$ is called a *canonical Gibbs measure* corresponding to the potential $\Phi$ if for all $\Lambda \in S$ and $\zeta \in \Omega_\Lambda$

$$\mu\,(X_\Lambda = \zeta \mid E_\Lambda) \quad = \quad \gamma_{\Lambda,N(X_\Lambda)}\,(\zeta\mid\cdot) \qquad \mu\text{ - a.s.}.$$

We let $C = C\,(\Phi)$ denote the set of all such canonical Gibbs measures.

(1.28) Remark: $C$ *is convex and weakly compact. For each* $z \in A$ *we have* $G\,(z) \subset C$ .

Proof: The compactness is an immediate consequence of the continuity of the functions $\gamma_{\Lambda,L}\,(\zeta\mid\cdot)$ which follows from (1.12) (ii) . The inclusion $G\,(z) \subset C$ is obtained from (1.25) . Indeed, for each $\mu \in G\,(z)$, $\Lambda \in S$, $L \in A_\Lambda$, $\zeta \in \Omega_{\Lambda,L}$, and $A \in F_{S\setminus\Lambda}$ we have

$$\int_{A\cap\{N(X_\Lambda)=L\}} 1_{\{X_\Lambda=\zeta\}}\,d\mu \quad = \quad \int_A \gamma^z_\Lambda\,(\zeta\mid\cdot)\,d\mu$$

$$= \int_A \gamma_{\Lambda,L}\,(\zeta\mid\cdot)\;\gamma^z_\Lambda\,(\Omega_{\Lambda,L}\mid\cdot)\,d\mu$$

$$= \int_{A\cap\{N(X_\Lambda)=L\}} \gamma_{\Lambda,L}\,(\zeta\mid\cdot)\,d\mu \quad .$$

The above remark implies that any superposition of measures in $\bigcup_z G\,(z)$ belongs to $C$ . In this text we will study the converse question: Under which conditions is every canonical Gibbs measure a superposition of Gibbs measures? Answers to this question would generalize the well-known theorem of B. de Finetti (1931) as is shown in the following

(1.29) Example: *Let* $\Phi \equiv 0$ . *Then* $C$ *consists of all symmetric probability measures on* $\Omega$ , *i.e., of all* $\mu$ *satisfying the condition*

$$\mu\,(X_\Lambda = \zeta) \quad = \quad \mu\,(X_\Lambda = \zeta')$$

*whenever* $\Lambda \in S$ *and* $\zeta'$ *is obtained from* $\zeta$ *by a permutation of the sites in* $\Lambda$ *(i.e.,* $N(\zeta) = N(\zeta')$*). Therefore de Finetti's theorem is equivalent to the statement: Any* $\mu \in C$ *is a unique mixture of products* $\alpha^S$ *of probability measures* $\alpha$ *on* $F$*, i.e., a unique mixture of Gibbs states, each corresponding to* $\Phi = 0$ *and some activity.*

Proof: For $\Phi = 0$ we have

$$\gamma_{\Lambda,L}(\zeta|\omega) = 1_{\Omega_{\Lambda,L}}(\zeta) / |\Omega_{\Lambda,L}|$$

and thereby for $\mu \in C$

$$\mu(X_\Lambda = \zeta) = \mu(N(X_\Lambda) = N(\zeta)) / |\Omega_{\Lambda,N(\zeta)}|$$

which shows that $\mu$ is symmetric. Conversely, if $\mu$ is symmetric it is easy to deduce that

$$\mu(X_\Lambda = \zeta | E_\Lambda) = \mu(X_\Lambda = \zeta' | E_\Lambda) \quad \text{a.s.}$$

whenever $\zeta, \zeta' \in \Omega_\Lambda$ and $N(\zeta) = N(\zeta')$. This proves that $\mu \in C$.

Various proofs of de Finetti's theorem can be found in Hewitt and Savage (1955) and the references therein, Feller (1966), Meyer (1966), and Dynkin (1978), e.g.. ┘

The following theorem gives us the main tools for the determination of the set $C$ in the general case: a description of the extreme points of $C$ and an integral representation of each element of $C$ by a probability measure on the extreme points. Recall that an element $\mu$ of a convex set $C$ is called an extreme point of $C$ if $\mu$ admits no representation of the form $\mu = \alpha\nu + (1-\alpha)\nu'$ with $0 < \alpha < 1$, $\nu, \nu' \in C$, $\nu \neq \nu'$. We let ex $C$ denote the set of extreme points of a convex set $C$.

Some more notations: For $\mu \in C$ we let

$$\Omega_\mu = \{\omega \in \Omega : \mu = \lim_{\Lambda \uparrow S} \gamma_{\Lambda,N(\omega_\Lambda)}(\cdot|\omega)\} \tag{1.30}$$

denote the set of all configurations $\omega$ such that the corresponding canonical Gibbs distributions converge weakly to $\mu$ ; more precisely, the set of all those $\omega$ for which

$$(1.31)\qquad \mu\,(X_V = \zeta) \;=\; \lim_{\Lambda\uparrow S} \gamma_{\Lambda,N(\omega_\Lambda)}\,(X_V = \zeta\,|\,\omega) \qquad (V \in S,\ \ \zeta \in \Omega_V)$$

holds.

(1.32) <u>Theorem:</u> (a) *$\mu \in C$ is extreme in $C$ iff $\mu(A) = 0$ or $1$ for all $A \in E_\infty$.*

(b) *Every $\mu \in \text{ex}\ C$ is the limit of canonical Gibbs distributions. More precisely: $\mu \in C$ is extreme in $C$ iff $\mu(\Omega_\mu) = 1$. In particular, any two different extreme points of $C$ are mutually singular.*

(c) *If $\text{ex}\ C$ is endowed with a suitable $\sigma$ - algebra then for every $\mu \in C$ there is a unique probability measure $P$ on $\text{ex}\ C$ such that*

$$\mu \;=\; \int_{\text{ex}\ C} \nu\ P\,(d\nu) \quad .$$

For a complete proof of this theorem see Preston (1976), Theorems 2.1 and 2.2 . We sketch the arguments. (a) follows immediately from the following easily verified fact: Suppose that $\mu \in C$ and that $f \geq 0$ is a measurable function satisfying $\int f\, d\mu = 1$ . Then $f\,\mu \in C$ iff $f$ has an $E_\infty$ - measurable version. The "only if" part of (b) follows from (a) by the martingale convergence theorem. (c) is a consequence of Choquet's theorem (It is readily seen that $C$ is even a Choquet simplex, see Lanford/Ruelle (1969) and Preston (1976), for instance). A more direct proof of (c) , which gives also the "if" part of (b) , is obtained by means of Martin-Dynkin-boundary techniques: The limits on the r.h.s. of (1.31) define (after some cosmetic operations) a probability kernel $K(\omega, A)$ $(\omega \in \Omega,\ A \in F)$ with the properties

(i) for all $\omega \in \Omega$ , $K(\omega, \cdot\,) \in \text{ex}\ C$ ,

(ii) for all $A \in F$, $K(\cdot, A)$ is $E_\infty$ - measurable and a version of $\mu(A|E_\infty)$ for each $\mu \in C$. In particular, any $\mu \in C$ has the representation

$$\mu = \int K(\omega, \cdot)\ \mu(d\omega) \quad . \tag{1.33}$$

Defining an equivalence relation on $\Omega$ by $\omega \sim \omega'$ iff $K(\omega, \cdot) = K(\omega', \cdot)$ and endowing the quotient space $\tilde{\Omega} = \Omega/\sim$ with the image $\tilde{E}_\infty$ of the $\sigma$ - algebra $E_\infty$ under the natural projection we obtain from (1.33) that any $\mu \in C$ has a unique representation

$$\mu = \int_{\tilde{\Omega}} K(\tilde{\omega}, \cdot)\ \tilde{\mu}(d\omega) \quad . \tag{1.33'}$$

Here $K(\tilde{\omega}, \cdot)$ denotes the value of the function $\omega \to K(\omega, \cdot)$ on the class $\tilde{\omega}$ and $\tilde{\mu}$ the image of $\mu|E_\infty$ on $(\tilde{\Omega}, \tilde{E}_\infty)$. Now it can be checked that the mapping $\tilde{\omega} \to K(\tilde{\omega}, \cdot)$ is a bijection from $\tilde{\Omega}$ onto ex $C$. Therefore, (c) is a consequence of (1.33').

This technique is due to Dynkin (for a survey see Dynkin (1978)) and was first applied to our situation by Föllmer (1975). For details see Preston (1976). All that is needed for this technique is the fact that $C$ is defined as a family of probability measures on a nice (standard Borel in sufficient) space whose conditional probabilities with respect to a decreasing family of sub - $\sigma$ - algebras are given by a consistent (in the sense of (1.26)) family of probability kernels. Since $G(z)$ is defined in a similar manner, we also have for any $z \in A$ :

(1.34) Theorem: *$\mu \in G(z)$ is extreme in $G(z)$ iff $\mu(A) = 0$ or $1$ for all $A \in F_\infty$, and this holds true iff for $\mu$ - a.e. $\omega$*

$$\mu = \lim_{\Lambda \uparrow S} \gamma_\Lambda^z(\cdot|\omega) \quad .$$

*Furthermore, for any $\mu \in G(z)$ there is a unique probability measure on* ex $G(z)$

*(with a suitable* $\sigma$ *- algebra) representing* $\mu$ .

We conclude this section with some remarks concerning the spatially homogeneous case. Suppose that $S = \mathbb{Z}^d$ for some $d \geq 1$ and that the potential $\Phi$ is shift-invariant in the sense of

$$(1.35) \qquad \Phi(A + x, \theta_x \omega) = \Phi(A, \omega) \quad (A \in \mathcal{S}, \ \omega \in \Omega, \ x \in S) \ ,$$

where $A + x = \{y + x : y \in A\}$ and the shift-group $\theta = (\theta_x)_{x \in S}$ on $\Omega$ is defined by

$$(1.36) \qquad (\theta_x \omega)_y = \omega_{y-x} \qquad (x, y \in S, \omega \in \Omega) \quad .$$

(1.35) implies that we always have

$$(1.37) \qquad \begin{aligned} \gamma_{\Lambda+x,L}(\theta_x \zeta | \theta_x \omega) &= \gamma_{\Lambda,L}(\zeta|\omega) \\ \gamma^z_{\Lambda+x}(\theta_x \zeta | \theta_x \omega) &= \gamma^z_\Lambda(\zeta|\omega) \end{aligned}$$

and therefore

$$(1.38) \qquad \begin{aligned} \theta_x(\mu) &= \mu(\theta_x^{-1}(.)) \in C \qquad \text{if} \quad \mu \in C \\ \theta_x(\mu) &\in G(z) \qquad \text{if} \quad \mu \in G(z) \quad . \end{aligned}$$

By the Markov-Kakutani fixed-point theorem this shows that both the convex and compact sets

$$(1.39) \qquad \begin{aligned} C_\theta &= \{\mu \in C : \theta_x(\mu) = \mu \ \text{ for all } \ x \in S\} \\ G_\theta(z) &= \{\mu \in G_\theta(z) : \theta_x(\mu) = \mu \ \text{ for all } \ x \in S\} \ , \end{aligned}$$

consisting of all shift-invariant canonical and grand canonical Gibbs measures, are non-empty. For the sets (1.39) one has results similar to (1.32) and (1.34) . Let us state only that the extreme points of $C_\theta$ and $G_\theta(z)$ are characterized by their ergodicity with respect to the shift-group and that every element of these sets is the barycentre of a unique probability measure on the extreme points, see Föllmer (1975) and Preston (1973, 1976) .

## 1.2 The continuous model

Let us start with $\mathbb{R}^d$ , the Euclidean space of dimension $d \geq 1$ , and $\sigma$ , a locally finite, positive measure on the Borel subsets of $\mathbb{R}^d$ . We assume that $\sigma$ is diffuse, i.e., that $\sigma(\{x\}) = 0$ for all $x \in \mathbb{R}^d$ . We will think of $\sigma$ as a quantity that specifies for each region of the space the tendency of the particles to remain there. In particular, the particles will only be allowed to occupy at most the sites of the closed support

(1.40) $$S = \{x \in \mathbb{R}^d : \sigma(U) > 0 \text{ for all open neighbourhoods } U \text{ of } x\}$$

of $\sigma$ . Therefore $S$ will be considered as the position space. We assume that $S$ is not bounded.

The most important special case is the homogeneous one when $\sigma$ is a multiple of Lebesgue measure $\lambda$ and $S = \mathbb{R}^d$ . Sometimes however (in barometric models, for instance), an inhomogeneous self-potential $\varphi : \mathbb{R}^d \to ]-\infty, \infty]$ is introduced; then we include $\varphi$ in $\sigma$ by putting $\sigma(dx) = e^{-\varphi(x)} \lambda(dx)$ .

We let $\mathcal{B}$ denote the $\sigma$ - algebra of the Borel subsets of $S$ and $\mathcal{S}$ the system of all bounded $\Lambda \in \mathcal{B}$ with $\sigma(\Lambda) > 0$ . In particular, $\mathcal{S}$ contains all bounded subsets of $\mathcal{S}$ which are open in $S$ . The notation $\Lambda \uparrow S$ will be used in the same sense as in (1.5) .

The most general configuration space which could be used is the set $M$ of all Radon measures on $S$ taking values only in the set $\{0, 1, 2, \dots, \infty\}$ . Besides the identically vanishing measure (which will be denoted by $\emptyset$ ) , $M$ consists of all measures of the form

(1.41) $$\omega = \sum_{i \geq 0} \varepsilon_{x_i} \quad ;$$

here $(x_i)_{i \geq 0}$ is a finite or infinite sequence in S which has no finite limit point, but may repeat the same point several times, and $\varepsilon_x$ denotes the Dirac measure corresponding to the point $x$. If $M$ is endowed with the vague topology, $M$ is a Polish space (see 4.1.2 and 4.1.3 in Kerstan/Matthes/Mecke (1974), for instance).

Obviously, any $\omega \in M$ can be thought of as a configuration of particles, several of which may have the same position. However, most of our work will be restricted to systems in equilibrium in which case it is natural to exclude multiple occupations. Therefore we chose the set

(1.42) $$\Omega = \{\omega : \omega(\{x\}) \leq 1 \text{ for all } x \in S\}$$

of all simple configurations as the configuration space. (Readers who feel the need to admit multiple configurations will see that this case could be included by a slight modification of our considerations below. In particular, we could then drop the requirement that $\sigma$ must be diffuse). As is well-known, $\Omega$ is a $G_\delta$ set in $M$ (see also (1.49) below).

Clearly, each $\omega \in \Omega$ is uniquely determined by its support, a locally finite subset of $S$. Therefore, henceforth the symbol $\omega$ will stand for both: a measure on $S$ and a subset of $S$. Correspondingly, we will be free to choose the more suggestive of equivalent notations as, for example, between $x \in \omega$ and $\omega(\{x\}) = 1$, $\{x\}$ and $\varepsilon_x$, $|\omega \cap \Lambda|$ and $\omega(\Lambda)$, $\sum_{x \in \omega} f(x)$ and $\int \omega(dx) f(x)$. Furthermore, we will use the shorthands $\omega\zeta$ instead of $\omega \cup \zeta$, in particular $x\zeta$ instead of $\{x\} \cup \zeta$, and $\omega_\Lambda$ instead of $\omega \cap \Lambda$.

We will consider also the case of particles with a hard core. To this end, we introduce a norm $\;\cdot\;$ on $R^d$ and think of the particles as rigid bodies with the shape of a ball $\{\;\cdot\; \leq r/2\}$ for some fixed $r > 0$. Then the only possible configurations are those which belong to the set

(1.43) $$\Omega_{(r)} = \{\omega \in \Omega : \|x - y\| > r \text{ for all } x, y \in \omega, x \neq y\} .$$

We will show in (1.49) that $\Omega_{(r)}$ is a $G_\delta$, too. Clearly, $\Omega = \Omega_{(0)}$.

The spaces $\Omega_{(r)}$, $r \geq 0$, contain some "dense" configurations which we want to exclude for reasons which will occur later. In $\Omega = \Omega_{(0)}$ we want to exclude those configurations $\omega$ whose particle number in outer regions increases so rapidly that

(1.44) $$\limsup_{\Lambda \uparrow S} \omega(\Lambda) / \sigma(\Lambda) = \infty .$$

In the case $r > 0$ we want to exclude those hard core configurations which make it impossible to add an additional particle even if we are allowed to push away the particles in a bounded region. In order to make this precise, we let

(1.45) $$M_\Lambda(\omega) = \max \{\zeta(\Lambda) : \zeta \in \Omega_{(r)}, \zeta_{S \setminus \Lambda} = \omega_{S \setminus \Lambda}\}$$

denote the largest number of balls that can be packed into the region $\Lambda \in \mathcal{S}$ when the balls outside of $\Lambda$ are fixed by $\omega \in \Omega_{(r)}$. Then a configuration $\omega \in \Omega_{(r)}$ will be considered as too densely packed if

(1.46) $$\omega(\Lambda) = M_\Lambda(\omega) \quad \text{for all} \quad \Lambda \in \mathcal{S} .$$

We will show at the end of section 6.3 that the phenomena (1.44) and (1.46) actually do not occur in some straightforward situations. Therefore it seems to be reasonable to exclude them from the beginning also in the general situation. Hence we fix a nice sequence $\Lambda_n \uparrow S$, say the sequence $\{x \in S : |x| < n\}, n \geq 1$, (where $| \cdot |$ denotes the Euclidean norm), and define

(1.47) $$\Omega_0 = \{ \omega \in \Omega : \limsup_{n \to \infty} \omega(\Lambda_n) / \sigma(\Lambda_n) < \infty \}$$

and for $r > 0$

(1.48) $$\Omega_r = \{ \omega \in \Omega_{(r)} : \omega(\Lambda) < M_\Lambda(\omega) \text{ for some } \Lambda \in S \} .$$

(1.49) Remark: *$\Omega_0$ is a $G_\delta \cap F_\sigma$ in $M$. For any $r > 0$, $\Omega_r$ and $\Omega_{(r)}$ are $G_\delta$ sets in $M$.*

Proof: First we observe that $\Omega_{(r)}$ is a $G_\delta$ for any $r \geq 0$. Indeed, for each compact $K \in S$ it is easily verified that the complement of

$$\Omega_{(r)}(K) = \{ \omega \in M : \omega((\,\cdot\,) \cap K) \in \Omega_{(r)} \}$$

is vaguely closed. $\Omega_{(r)}$ is the intersection of those $\Omega_{(r)}(K)$ when $K$ runs through a cofinal sequence in $S$. In particular, $\Omega_0$ is the intersection of the $G_\delta$ set $\Omega = \Omega_{(0)}$ and the $F_\sigma$ set

$$\{ \omega \in M : \limsup_{n \to \infty} \omega(\Lambda_n) / \sigma(\Lambda_n) < \infty \} .$$

In order to see that $\Omega_r$ is a $G_\delta$ for all $r > 0$, note that $\Omega_r$ is the intersection of $\Omega_{(r)}$ and the union of all sets of the form

$$G = \{ \omega \in \Omega_{(r)}(K) : \omega(\Lambda) < M_\Lambda(\omega) \} ,$$

where $\Lambda \in S$ and $K$ is a compact neighbourhood of $\Lambda$ whose complement has a $\| \cdot \|$-distance greater than $2r$ from $\Lambda$. Thus we must only show that such a $G$ is open. Choose any $\omega \in G$ and let $\zeta \in \Omega_{(r)}$ be such that $\zeta(\Lambda) > \omega(\Lambda)$ and $\zeta_{K \setminus \Lambda} = \omega_{K \setminus \Lambda}$. Let $0 < \varepsilon < r/2$ be so small that

$$2\varepsilon < \min \{ \| x - y \| : x, y \in \zeta_K, x \neq y \} - r .$$

Now choose a continuous function $f$ with the properties (i) $0 \leq f \leq 1$, (ii) $f(x) = 0$ for all $x \in \omega_K \cup (S \setminus K)$, and (iii) $f(y) = 1$ whenever $y \in K$ and $\| y - x \| \geq \varepsilon$ for all $x \in \omega_K \cup (S \setminus K)$. Then

$$U \quad = \quad \{ \eta \in \Omega_{(r)} (K) : |\int f \, d\eta - \int f \, d\omega| < 1 \}$$

is an open neighbourhood of $\omega$ which is contained in $G$ . Indeed, if $\eta \in U$ then every $x \in \eta_K$ is close either to $S \setminus K$ or to some $x' \in \omega_K$ , and by a slight modification of $\omega_{K \setminus \Lambda} = \zeta_{K \setminus \Lambda}$ we obtain from $\zeta$ a configuration $\zeta'$ satisfying $\zeta'_{K \setminus \Lambda} = \eta_{K \setminus \Lambda}$ and $\zeta'(\Lambda) > \eta(\Lambda)$ . ┘

For any $V \in \mathcal{B}$ we let $N(V)$ denote the counting variable corresponding to $V$ which is defined by

(1.50) $$N (V) (\omega) \quad = \quad \omega (V) \qquad (\omega \in M) ,$$

and we let

(1.51) $$\mathcal{F}_V \quad = \quad \sigma (N (\Lambda) : V \supset \Lambda \in \mathcal{S})$$

denote the $\sigma$ - algebra of the events in $V$ . It will be clear from the context whether $\mathcal{F}_V$ is understood as a $\sigma$ - algebra on either $M$, $\Omega$, or $\Omega_r$, $r \geq 0$ . It is well-known that $\mathcal{F} = \mathcal{F}_S$ on $M$ coincides with the $\sigma$ - algebra which is generated by the vague topology, see 4.1.5 in Kerstan/Matthes/Mecke (1974) , for instance. Therefore (1.49) implies that the measurable spaces $(\Omega_r, \mathcal{F})$ , $r \geq 0$ , are standard Borel spaces. As in the discrete case, we will also consider the tail field

(1.52) $$\mathcal{F}_\infty \quad = \quad \bigcap_{\Lambda \in \mathcal{S}} \mathcal{F}_{S \setminus \Lambda}$$

and the $\sigma$ - field of symmetric events

(1.53) $$\mathcal{E}_\infty \quad = \quad \bigcap_{\Lambda \in \mathcal{S}} \mathcal{E}_\Lambda \quad .$$

Here for any $\Lambda \in \mathcal{S}$

(1.54) $$\mathcal{E}_\Lambda \quad = \quad \sigma (N (\Lambda) , \mathcal{F}_{S \setminus \Lambda})$$

denotes the $\sigma$ - algebra of all events which are invariant under rearrangements of the particles in $\Lambda$ .

As stated above, the preference of the particles for the different regions of the space will be specified by the measure $\sigma$ . This means that the mean particle number in $\Lambda \in S$ should be $\sigma(\Lambda)$ if there is no interaction of the particles to prevent this. Therefore an ideal gas, i.e., a system of non-interacting particles, should satisfy the following requirement: For any pairwise disjoint bounded Borel sets $\Lambda_1,\ldots,\Lambda_n$ , the random variables $N(\Lambda_1),\ldots,N(\Lambda_n)$ on $\Omega$ are independent with corresponding expected values $\sigma(\Lambda_1),\ldots,\sigma(\Lambda_n)$ . Since $\sigma$ is assumed to be diffuse, there is one (and only one) probability measure on $(\Omega, F)$ with this property, namely the *Poisson point process* $\pi$ *with intensity measure* $\sigma$ . Its restriction to $F_\Lambda$ for a bounded set $\Lambda \in B$ is defined as follows: For any $F_\Lambda$ - measurable function $f \geq 0$ the equality

$$(1.55) \qquad \int \pi\,(d\omega)\; f\,(\omega) \;=\; e^{-\sigma(\Lambda)} \sum_{n\geq 0} \frac{1}{n!} \int_{\Lambda^n} \sigma(dx_1),\ldots,\sigma(dx_n)\; f\,(\{x_1,\ldots,x_n\})$$

holds. We will often use the fact that for any two disjoint Borel sets $V$ and $W$ the $\sigma$ - algebras $F_V$ and $F_W$ are independent with respect to $\pi$ , i.e., that for any nonnegative $F_{V\cup W}$ - measurable function $f$ the equality

$$(1.56) \qquad \int \pi\,(d\omega)\; f\,(\omega) \;=\; \int \pi\,(d\zeta) \int \pi\,(d\eta)\; f\,(\zeta_V \eta_W)$$

is true. A further useful property of $\pi$ is the following:

$$(1.57) \qquad \int \pi\,(d\omega) \int \omega\,(dx)\; g\,(x,\omega) \;=\; \int \sigma\,(dx) \int \pi\,(d\omega)\; g\,(x,\, x\omega)$$

whenever $g$ is a nonnegative $B \otimes F$ -measurable function on $S \times \Omega$ . It is easy to deduce this fact from formula (1.55). Actually, $\pi$ is characterized by property (1.57) ; this result which is due to Mecke has been extended to Gibbs measures by Nguyen/Zessin (1976) . Finally, we note that a version of the law of large numbers shows that $\pi\,(\Omega_0) = 1$ , see 1.6.5 in Kerstan/Matthes/Mecke (1974) or (4.29) below.

Now we introduce an interaction for the particles. Let

$$\Omega_f \quad = \quad \{ \alpha \in \Omega : 1 \leq \alpha(S) < \infty \} \quad .$$

(1.58) <u>Definition:</u> An F - measurable function

$$\Phi : \Omega_f \longrightarrow ]-\infty , \infty]$$

is called a *potential* if $\Phi$ has the following properties:

(i) There is a particle diameter $r = r(\Phi) \geq 0$ such that $\Phi(\alpha) < \infty$ for all $\alpha \in \Omega_f \cap \Omega_{(r)}$ and $\Phi(\{x, y\}) = \infty$ whenever $0 < \| x - y \| \leq r$ .

(ii) The *energy* of $\zeta \in \Omega$ in $\Lambda \in S$ with the *boundary condition* $\omega \in \Omega_{(r)}$ given by

$$E_\Lambda(\zeta|\omega) \quad = \quad \sum_{\alpha \subset \zeta_\Lambda \omega_{S \setminus \Lambda}, \alpha(\Lambda) > 0} \Phi(\alpha)$$

is always well-defined (as the finite or infinite limit of the partial sum over all $\alpha \subset V$ when $V$ runs through the directed set $S$ ) and is finite iff $\zeta_\Lambda \omega_{S \setminus \Lambda} \in \Omega_{(r)}$ .

(iii) For each $\Lambda \in S$ and $\omega \in \Omega_{(r)}$ there exists a number $B < \infty$ such that

$$E_\Lambda(\zeta|\omega) \geq - B \, \zeta(\Lambda)$$

for all $\zeta \in \Omega$ .

(iv) For all $\Lambda \in S$ and $\omega , \zeta \in \Omega$ with $\zeta_\Lambda \omega_{S \setminus \Lambda} \in \Omega_{(r)}$ ,

$$\sup \{ |E_\Lambda(\zeta|\omega) - E_\Lambda(\zeta|\omega')| : \omega' \in \Omega_{(r)}, \omega'_{V \setminus \Lambda} = \omega_{V \setminus \Lambda} \}$$

tends to zero in the limit $V \uparrow S$ .

If $\Phi$ has a hard core (i.e. $r(\Phi) > 0$ ) then the requirements (ii) - (iv) are as natural as in the discrete case: They are satisfied whenever $\Phi(\alpha)$ decreases rapidly enough as diam $\alpha$ increases. In the case $r(\Phi) = 0$ however, the conditions (ii) - (iv) are rather restrictive; they are satisfied iff $\Phi$ has finite range, i.e., iff for any $\Lambda \in S$ there is some $\Delta \in S$ such that $\Lambda \subset \Delta$ and $\Phi(\alpha) = 0$

whenever $\alpha(\Lambda)\,\alpha(S \setminus \Delta) > 0$ .

Often we will consider the energy which is required in order to add an extra particle at the site $x \in S$ to a configuration $\omega \in \Omega_{(r)}$ , namely

$$(1.59) \qquad E(x|\omega) = \sum_{x \in \alpha \subset x\omega} \Phi(\alpha) \quad .$$

Of particular interest are *pair potentials*, i.e., potentials $\Phi$ such that $\Phi(\alpha) = 0$ whenever $\alpha(S) > 2$ . For these we use the shorthand

$$(1.60) \qquad \Phi(x, y) = \Phi(\{x, y\}) \qquad (x, y \in S) \quad .$$

Thus a pair potential is a $\mathcal{B} \otimes \mathcal{B}$ - measurable symmetric function $\Phi : S \times S \to ]-\infty, \infty]$ satisfying $\Phi(x, x) = 0 \;\; (x \in S)$ , $\Phi(x, y) = \infty$ iff $0 < \| x - y \| \leq r(\Phi)$ , and conditions (ii) - (iv) .

As in (1.15) , we need a parameter which controls the particle density. This parameter is again called the *activity* and is (since we have only one type of particle) just a real number $z \geq 0$ .

From now on, we suppose that a fixed potential $\Phi$ is given. Together with an activity $z \geq 0$ , $\Phi$ defines a system of conditional probabilities as follows:

(1.61) <u>Definition:</u> Suppose that $\Lambda \in \mathcal{S}$ and $\omega \in \Omega_{r(\Phi)}$ . Then the probability measure on $(\Omega, \mathcal{F}_\Lambda)$ whose Radon-Nikodym density with respect to $\pi|\mathcal{F}_\Lambda$ is given by

$$\gamma_\Lambda^z(\zeta|\omega) = Z_\Lambda(z, \omega)^{-1} \; z^{\zeta(\Lambda)} \; \exp[-E_\Lambda(\zeta|\omega)] \qquad (\zeta \in \Omega)$$

is called the *(grand canonical) Gibbs distribution* on $\Lambda$ with boundary condition $\omega$ corresponding to the potential $\Phi$ and the activity $z$ . The normalizing factor

$$Z_\Lambda(z, \omega) = \int \pi(d\zeta) \; z^{\zeta(\Lambda)} \; \exp[-E_\Lambda(\zeta|\omega)]$$

is called the *partition function*.

Notice that $\gamma_\Lambda^z(\cdot \mid \omega)$ is well-defined since $0 < Z_\Lambda(z, \omega) < \infty$ . The positivity of the partition function comes from the fact that $\pi(N(\Lambda) = 0) = e^{-\sigma(\Lambda)} > 0$ and $E_\Lambda(\cdot \mid \omega) = 0$ on $\{N(\Lambda) = 0\}$ . The stability condition (1.58) (iii) guarantees that the partition function is finite since

$$\int \pi(d\zeta) \ (z\, e^B)^{\zeta(\Lambda)} = \exp[\sigma(\Lambda)(z\, e^B - 1)] < \infty \quad .$$

It is easily verified that the grand canonical densities $\gamma_\Lambda^z$ satisfy a consistency condition similar to (1.17) . For the converse question of whether a system of conditional densities can be described in terms of a potential we refer to Kozlov (1976) and Glötzl (1978) .

Now we are going to define Gibbs measures as those measures whose conditional probabilities with respect to $F_{S \setminus \Lambda}$ are given by the Gibbs distributions in $\Lambda$ . It is convenient to formulate this fact by means of the measure $\tilde{\mu}$ on $(\Omega \times \Omega, F \otimes F)$ which is obtained by putting $\mu$ on the diagonal of $\Omega \times \Omega$ and is defined by

(1.62) $$\tilde{\mu}(A \times B) = \mu(A \cap B) \qquad (A, B \in F) \quad .$$

Indeed, the statement that for any $A \in F_\Lambda$

$$\int_A \pi(d\zeta) \ \gamma_\Lambda^z(\zeta \mid \cdot \,)$$

is a version of $\mu(A \mid F_{S \setminus \Lambda})$ is true iff $\gamma_\Lambda^z(\cdot \mid \cdot)$ is a Radon-Nikodym density of $\tilde{\mu}$ with respect to $\pi \otimes \mu$ on $F_\Lambda \otimes F_{S \setminus \Lambda}$ .

(1.63) Definition: A probability measure $\mu$ on $(\Omega, F)$ is called a *Gibbs measure with respect to the activity* $z \geq 0$ and the potential $\Phi$ if $\mu(\Omega_{r(\Phi)}) = 1$ and for all $\Lambda \in S$

$$d\tilde{\mu} / d(\pi \otimes \mu) \mid F_\Lambda \otimes F_{S \setminus \Lambda} = \gamma_\Lambda^z(\cdot \mid \cdot) \qquad \pi \otimes \mu \text{ - a.s.}$$

The convex set of all such Gibbs measures is denoted by $G(z) = G(z, \Phi)$ .

In the continuous model it is much more difficult than in the discrete model to prove the existence of Gibbs measures since in the former case the set of all probability measures on $\Omega$ is not compact. Therefore it is necessary to enforce the compactness of suitable sets of probability measures by making appropriate assumptions on the potential $\Phi$ . This can easily be done in the case of a hard core, see Dobrushin (1969) , but is intricate if $r(\Phi) = 0$ ; we refer to Dobrushin (1970) , Ruelle (1970) , and Preston (1976) . Here we write down only the following vague statement.

(1.64) <u>Theorem:</u> *There are some potentials $\Phi$ for which $G(z)$ is non-empty for all $z \geq 0$ . Sometimes $G(z)$ , or at least a reasonable subset of $G(z)$ , is sequentially compact in a suitable topology.*

In some trivial cases, $G(z)$ is easily identified:

(1.65) <u>Example:</u> *If $\Phi \equiv 0$ then for any $z \geq 0$ $G(z)$ is a singleton consisting of the Poisson process $\pi^z$ corresponding to the intensity measure $z\,\sigma$ . For any $\Phi$ we have $G(0) = \{\varepsilon_\emptyset\}$ where $\varepsilon_\emptyset$ denotes the Dirac measure on the empty configuration $\emptyset$ .*

Now we introduce the corresponding canonical notions.

(1.66) <u>Definition:</u> Suppose that $\Lambda \in S$ , $N \geq 0$ , and $\omega \in \Omega_{r(\emptyset)}$ . Then the measure on $(\Omega, F_\Lambda)$ whose density with respect to $\pi | F_\Lambda$ is given by

$$\gamma_{\Lambda,N}(\zeta|\omega) = \begin{cases} Z_{\Lambda,N}(\omega)^{-1} \, 1_{\{N(\Lambda)=N\}}(\zeta) \, \exp[-E_\Lambda(\zeta|\omega)] & \text{if } Z_{\Lambda,N}(\omega) > 0 \\ 0 & \text{otherwise} \end{cases}$$

is called the *canonical Gibbs distribution* on $\Lambda$ for $\Phi$ corresponding to the particle number $N$ and the boundary condition $\omega_{S\setminus\Lambda}$ . The normalizing factor

$$Z_{\Lambda,N}(\omega) = \int_{\{N(\Lambda)=N\}} \pi(d\zeta) \exp[-E_\Lambda(\zeta|\omega)]$$

is called the *partition function* for the particle number $N$.

The condition (1.58) (iii) on $\Phi$ guarantees that $Z_{\Lambda,N}(\omega) < \infty$. Furthermore, we have $Z_{\Lambda,N}(\omega) > 0$ whenever $r(\Phi) = 0$. In the case of a hard core $r(\Phi)>0$, $Z_{\Lambda,N}(\omega)$ vanishes for any $N > M_\Lambda(\omega)$; however, the inequality $Z_{\Lambda,\omega(\Lambda)}(\omega) > 0$ holds whenever $\omega \in \Omega_{r(\Phi)}$ and $\sigma(U \cap \Lambda) > 0$ for each neighbourhood $U$ of any $x \in \omega_\Lambda$. This is because the latter condition (which of course is satisfied if $\Lambda$ is open in $S$) implies that the set of all configurations $\zeta$, which on $\Lambda$ are close enough to $\omega_\Lambda$ in order to satisfy $E_\Lambda(\zeta|\omega) < \infty$, has positive $\pi$ - measure. If $N < 0$ we put $Z_{\Lambda,N}(\omega) = 0$.

It is not difficult to see that the canonical densities $\gamma_{\Lambda,\omega(\Lambda)}(\zeta|\omega)$ satisfy a consistency condition similar to (1.26). Moreover, they are compatible with the grand canonical densities in the sense that

$$(1.67) \qquad \gamma_\Lambda^z(\zeta|\omega) = \gamma_{\Lambda,\omega(\Lambda)}(\zeta|\omega) \int_{\{N(\Lambda)=\omega(\Lambda)\}} \pi(d\eta)\, \gamma_\Lambda^z(\eta|\omega)$$

whenever $\Lambda \in S$, $z \geq 0$, $\omega \in \Omega_{r(\Phi)}$, and $\zeta \in \Omega$ with $\zeta(\Lambda) = \omega(\Lambda)$. This is a consequence of (1.61) and (1.66).

Finally, we observe that the canonical and grand canonical Gibbs distributions remain unchanged if $\sigma$ is replaced by $e^{-\Phi(\{x\})}\sigma(dx)$ and $\Phi$ by $1_{\{N(S)>1\}}\Phi$. Therefore, without loss of generality we can (and will) assume throughout that the self-potential $x \to \Phi(\{x\})$ vanishes. The only exception will be made in section 2.2 when the potential also is used to define certain time evolutions.

(1.68) <u>Definition:</u> Suppose that $\mu$ is a probability measure on $(\Omega, F)$ such that $\mu(\Omega_{r(\Phi)}) = 1$. $\mu$ is called a *canonical Gibbs measure* if for all $\Lambda \in S$ the function

$$(\zeta, \omega) \to \gamma_{\Lambda,\omega(\Lambda)}(\zeta|\omega)$$

is a Radon-Nikodym density of $\tilde{\mu}$ with respect to $\pi \otimes \mu$ on the $\sigma$ - algebra $F_\Lambda \otimes E_\Lambda$ . We let $C = C(\Phi)$ denote the convex set of all such canonical Gibbs measures.

Notice that each $\mu \in C$ is by definition locally absolutely continuous with respect to $\pi$ . Indeed, for any $\Lambda \in S$

$$(1.69) \qquad u_\Lambda^\mu = \int \mu\,(d\omega)\ \gamma_{\Lambda,\omega(\Lambda)}\,(\,\cdot\,|\,\omega)$$

is a density of $\mu | F_\Lambda$ with respect to $\pi | F_\Lambda$ . Moreover, in the hard core case it is useful to notice that for all $\mu \in C$ and $\Lambda \in S$ we have

$$(1.70) \qquad \mu\,(\omega \in \Omega_{r(\Phi)} \ :\ Z_{\Lambda,\omega(\Lambda)}\,(\omega) > 0) = 1 \quad ,$$

since this probability is given by

$$\int \mu\,(d\omega) \int \pi\,(d\zeta)\ \gamma_{\Lambda,\omega(\Lambda)}\,(\zeta\,|\,\omega) = \mu\,(\Omega) \ .$$

As an immediate consequence of (1.67) we obtain just as in (1.28) :

(1.71) Remark: *For any* $z \geq 0$ *we have* $G(z) \subset C$ . *In particular* $\varepsilon_\emptyset \in \mathrm{ex}\ C$ .

We see from example (1.65) that $C$ will not in general be compact in any natural topology. Nevertheless we can represent any $\mu \in C$ by a probability measure on ex $C$ since the Martin-Dynkin boundary technique (explained after (1.32)) needs only the fact that the spaces $(\Omega_r, F)$ , $r \geq 0$ , are standard Borel and we have already seen that this is true. This gives part (b) of :

(1.72) Theorem: (a) $\mu \in C$ *is extreme in* $C$ *iff* $\mu$ *satisfies a* $0 - 1$ *law on* $E_\infty$ .
(b) *If* ex $C$ *is endowed with a suitable* $\sigma$ *- algebra then for any* $\mu \in C$ *there is one and only one probability measure on* ex $C$ *with barycentre* $\mu$ .

Furthermore, we have an approximation statement similar to (1.32) (b) . We want to make precise which kind of convergence is used in this approximation. To this end, we define for any $V, \Lambda \in S$ with $V \subset \Lambda$ and all $\zeta \in \Omega$, $\omega \in \Omega_{r(\Phi)}$

(1.73) $$\gamma_{V\Lambda,\omega(\Lambda)}(\zeta|\omega) = \int \pi(d\eta)\, \gamma_{\Lambda,\omega(\Lambda)}(\zeta_{V\cap S\setminus V}|\omega) \quad .$$

Then we have:

(1.74) Corollary: *Suppose that* $\eta \in \text{ex } C$ *and* $V \in S$ . *Then for* $\mu$ - a.e.$\omega$ *the sequence* $\gamma_{V\Lambda,\omega(\Lambda)}(\,\cdot\,|\omega)$ *in the limit* $\Lambda \uparrow S$ *tends to* $u_V^\mu = d\mu/d\pi \mid F_V$ $\pi$ *- a.s. and in* $L^1(\pi)$ *- norm. In particular, any two different extreme points of* $C$ *are mutually singular.*

Proof: Clearly, the function $(\zeta,\omega) \to \gamma_{V\Lambda,\omega(\Lambda)}(\zeta|\omega)$ is a Radon-Nikodym density of $\tilde{\mu}$ with respect to $\pi \otimes \mu$ on $F_V \otimes E_\Lambda$ . Thus the martingale convergence theorem implies that

$$u_V(\zeta|\omega) = \lim_{\Lambda\uparrow S} \gamma_{V\Lambda,\omega(\Lambda)}(\zeta|\omega)$$

exists $\pi \otimes \mu$ - a.s. and in $L^1(\pi \otimes \mu)$ - norm. In order to identify the limit $u_V$ notice that because of (1.72) (a) the sequence

$$\int_A \pi(d\zeta)\ \gamma_{V\Lambda,\omega(\Lambda)}(\zeta|\,\cdot\,) = \mu(A|E_\Lambda)$$

converges in $L^1(\mu)$ - norm to $\mu(A) = \int_A u_V^\mu\, d\pi$ for any $A \in F_V$ . This shows that $u_V(\zeta|\omega) = u_V^\mu(\zeta)$ for $\pi \otimes \mu$ - a.e. $(\zeta, \omega)$ , and the corollary follows. ┘

Similar statements to (1.72) and (1.74) are true for the set $G(z)$ of Gibbs measures with respect to an activity $z \geq 0$ - see the discrete version (1.34). As a result of particular interest, the Martin-Dynkin boundary technique gives: If $G(z) \neq \emptyset$ then $\text{ex } G(z) \neq \emptyset$ .

For the spatially homogeneous case $\sigma = \lambda$ , $S = R^d$, we will use the notations (1.36) and (1.39) also in the continuous model. It is known that for some shift - invariant potentials $\Phi$ the set $G_\theta(z)$ (and therefore also ex $G_\theta(z)$) is non-empty for all $z > 0$ (see the references just before (1.64) ) and that in general the extreme points in $G_\theta(z)$ and $C_\theta$ are always characterized by their ergodicity with respect to the shift-group.

*Bibliographical notes:* Canonical Gibbs states first appeared in the papers of R. Holley (1971) and K.G. Logan (1974), but in a somewhat different form. R.L. Thompson (1974) was the first to study this notion in its own right. Independently, H. Rost suggested considering canonical Gibbs states; this led to the papers Georgii (1975, 1976). Again independently, M. Aizenman et al. (1977) were led to introduce this concept.

## § 2 Equilibrium states for systems of moving particles

In this chapter we will explain from where the interest in canonical Gibbs measures comes. In a physical system each of the particles is forced by the interaction to arrange its speed so that the energy of the configuration in its neighbourhood decreases as rapidly as possible. We will consider several models for a motion of this type and will see that, roughly speaking, canonical Gibbs measures are those measures which are not only globally but also locally in equilibrium.

### 2.1 Particle motions in the discrete model

We are going to construct a time evolution for the discrete model which was introduced in section 1.1 . Suppose that for each two-point set $xy = \{x, y\}$ $(x \neq y)$ in $S$ and any $\omega \in \Omega$ there is given a nonnegative number $c(xy, \omega)$ which will be the rate at which the particles at the sites $x$ and $y$ interchange their positions when the total configuration is $\omega$ . We let ${}^{xy}\omega$ denote the configuration after this interchange, i.e.,

$$(2.1) \qquad ({}^{xy}\omega)_u = \begin{cases} \omega_u & \\ \omega_x & \text{if} \\ \omega_y & \end{cases} \begin{array}{l} u \neq x, y \\ u = y \\ u = x \end{array} .$$

Since an interchange of particles of equal type does not change the configuration we can (and will) assume that

$$(2.2) \qquad c\,(xy, \omega) = 0 \quad \text{if} \quad \omega_x = \omega_y .$$

A Markov process $((\xi_t)_{t\geq 0}\, ,\, (P^\omega)_{\omega\in\Omega})$ with state space $\Omega$ (and stationary transition probabilities) will be called a *particle jump process* with rates $c\,(\cdot, \cdot)$ if the condition

$$P^{\omega}(\xi_t = \zeta) = \begin{cases} c(xy,\omega)\, t + o(t) & \text{if} \quad \zeta = {}^{xy}\omega,\ \omega_x \neq \omega_y \\ \\ o(t) & \text{if} \quad \zeta \neq {}^{xy}\omega \ \text{ for all } \ xy \end{cases}$$

is satisfied. It is well-known that in order to prove the existence of such a process it is sufficient to find a Markov semigroup $(P_t)_{t\geq 0}$ on the space $C(\Omega)$ of all continuous real functions on $\Omega$, the generator of which is determined by the requirement above. If $f$ is a function on $\Omega$ which depends only on finitely many coordinates (we let $C_0(\Omega)$ denote the set of these so-called cylinder functions) then we obtain from the above requirement the following form for the generator:

$$(2.3) \qquad Gf(\omega) = \sum_{xy \subset S} c(xy,\omega)\,\{f({}^{xy}\omega) - f(\omega)\} \quad .$$

The theorem below which is due to Liggett (1971) ensures the existence and uniqueness of a particle jump process with given rates provided that these are, in a certain sense, local. For the proof we refer to Liggett (1971, 1977) and Sullivan (1975) .

(2.4) Theorem: *For any two-point set* $xy \subset S$ *let* $c(xy, \cdot\,)$ *be a nonnegative continuous function on* $\Omega$ *satisfying* (2.2) . *Suppose that*

$$\sup_{x\in S} \sum_{y\neq x} \sup_{\omega\in\Omega} c(xy,\omega) < \infty$$

*and*

$$\sup_{x\in S} \sum_{y\neq x} \sum_{u\in S} \sup_{\substack{\omega,\zeta\in\Omega \\ \omega_v = \zeta_v \text{ unless } v = u}} |c(xy,\omega) - c(xy,\zeta)| < \infty$$

*Then the closure* $\overline{G}$ *of the operator* $(G, C_0(\Omega))$ *generates a unique Markov semigroup* $(P_t)_{t\geq 0}$ *on* $\Omega$ .

In the sequel we will assume that the rate function $c(\cdot\,,\cdot)$ under consideration satisfies the conditions of Theorem (2.4) .

Now we discuss some special choices for the rate function $c(\cdot,\cdot)$ . For any $a \in F$ let $p_a = (p_a(x,y))_{x,y\in S}$ be a nonnegative (for example, stochastic) matrix. We think of $p_a(x,y)$ as the rate at which a particle of type a would jump from x to y if there were no other particles. This rate is changed by the interaction with the other particles (which is given by a potential $\Phi$ ). The modification could depend on either the common energy

$$(2.5) \qquad E(xy, \omega) = E_{\{x,y\}}(\omega_x\omega_y|\omega)$$

of the particles at x and y or on the value of the energy gain $E(xy, {}^{xy}\omega) - E(xy, \omega)$ resulting from the jump. Particular choices are:

$$(2.6) \qquad c(xy, \omega) = [p_{\omega_x}(x,y) + p_{\omega_y}(y,x)] \exp E(xy, \omega)$$

$$(2.7) \qquad c(xy, \omega) = [p_{\omega_x}(x,y) + p_{\omega_y}(y,x)] \exp [\tfrac{1}{2} E(xy, \omega) - \tfrac{1}{2} E(xy, {}^{xy}\omega)]$$

$$(2.8) \qquad c(xy, \omega) = [p_{\omega_x}(x,y) + p_{\omega_y}(y,x)] / \{1 + \exp [E(xy, {}^{xy}\omega) - E(xy, \omega)]\}$$

whenever $\omega_x \neq \omega_y$ . In the special case when $F = \{0, 1\}$ , $p_0 \equiv 0$ , $p_1 = p$ a stochastic matrix, and $\Phi$ is of the normalized form (1.14) , the rate (2.6) reduces to

$$(2.9) \quad c(xy, \omega) = \omega_x(1 - \omega_y)\, p(x,y) \exp E_x(1|\omega) + \omega_y(1 - \omega_x)\, p(y,x) \exp E_y(1|\omega) .$$

The rate (2.9) defines a "speed change with exclusion" process as introduced by Spitzer (1970) . Its intuitive description is that a particle at a site x decides with a rate $\exp E_x(1|\omega)$ to change its position and then chooses its destination y according to p ; if y is already occupied the particle decides to stay at x . The rate (2.8) has been used by Bortz et al. for the description of the time evolution of a binary alloy; for references see Liggett (1977) .

It is easily verified that the rate functions (2.6), (2.7), and (2.8) satisfy the conditions of Theorem (2.4) if

$$\sup_{x} \sum_{y \neq x} \max_{a \neq b} [ p_a(x,y) + p_b(y,x) ] < \infty$$

and

$$\sup_{x} \sum_{A \ni x} |A| \, \| \Phi(A, \cdot) \| < \infty .$$

The intuitive notion of being "locally in equilibrium", which we used at the beginning of this chapter, is made precise by the concept of reversibility.

(2.10) <u>Definition:</u> A probability measure $\mu$ on $(\Omega, F)$ is said to be *reversible* for a Markov process with semigroup $(P_t)_{t \geq 0}$ acting on $C(\Omega)$ if for all $f, g \in C(\Omega)$ and $t > 0$ the equation

$$\int f \, P_t \, g \; d\mu = \int g \, P_t \, f \; d\mu$$

holds.

It is well-known that $\mu$ is reversible in the sense of this definition iff for all $t > 0$ the corresponding Markov process $(\xi_s)_{0 \leq s \leq t}$ with initial distribution $\mu$ is equivalent to $(\xi_{t-s})_{0 \leq s \leq t}$. We let $R(c)$ denote the set of all probability measures which are reversible for the particle jump process with rate function $c(\cdot\, , \cdot)$. If $R(c)$ is non-empty then it is convex and weakly compact.

Suppose now that we are given a potential $\Phi$. We ask for conditions on $c(\cdot\, , \cdot)$ which imply that $R(c)$ contains a Gibbs measure with respect to $\Phi$ and a non-degenerate activity. The answer is given by the next remark which we will prove in (2.25).

(2.11) <u>Remark:</u> *Suppose that* $R(c) \cap G(z, \Phi) \neq \emptyset$ *for some* $z \in A$ *with* $\prod_{a \in F} z(a) > 0$. *Then* $\Phi$ *and* $c(\cdot\, , \cdot)$ *are coupled by the following relation* $(R_\Phi)$.

$(R_\Phi)$ *For any* $xy \subset S$ *the function* $d(xy, \cdot) = c(xy, \cdot) \exp[- E(xy, \cdot)]$ *is measurable with respect to* $E_{xy}$.

In simpler terms, $(R_\Phi)$ just means that

$$(2.12) \qquad c(xy,\omega)\, e^{-E(xy,\omega)} = c(xy, {}^{xy}\omega)\, e^{-E(xy,\, {}^{xy}\omega)}$$

whenever $xy \subset S$ and $\omega \in \Omega$.

We will prove below that $(R_\Phi)$ in turn implies that $C(\Phi) \subset R(c)$. But first let us check whether the rate functions (2.6), (2.7), and (2.8) satisfy condition $(R_\Phi)$. Clearly, this is the case when for any $a \in F$ the matrix $p_a$ is symmetric. Indeed, for the rate (2.6) we have $d(xy,\omega) = p_{\omega_x}(x,y) + p_{\omega_y}(y,x)$, and in the cases (2.7) and (2.8) we get the additional factors

$$\exp\left[-\tfrac{1}{2} E(xy,\omega) - \tfrac{1}{2} E(xy, {}^{xy}\omega)\right]$$

and

$$\{ \exp E(xy,\omega) + \exp E(xy, {}^{xy}\omega) \}^{-1} ,$$

respectively. However, the symmetry of the matrices $p_a$ is not necessary in order to find a potential $\Psi$ (which may be different from $\Phi$) such that $(R_\Psi)$ is satisfied by the rates (2.6 - 8). For example, suppose that $p$ is a nonnegative matrix such that the "detailed balance" equation

$$(2.13) \qquad \sigma(x)\, p(x,y) = \sigma(y)\, p(y,x) \qquad (x,y \in S)$$

has a strictly positive solution $\sigma$. Assume that for any $a \in F$ either $p_a \equiv 0$ or $p_a = p$ (as, for instance, in Spitzer's case (2.9) ). Then a short calculation shows that the rates (2.6 - 8) satisfy $(R_\Psi)$ if the potential $\Psi$ is given by

$$\Psi(A,\omega) = \begin{cases} \Phi(x,\omega) - \log \sigma(x) & \text{if } A = \{x\},\ p_{\omega_x} = p \\ \Phi(A,\omega) & \text{otherwise} . \end{cases}$$

It is natural to require that each particle can eventually reach every site. Thus we call a rate function $c(\cdot\,,\cdot)$ *irreducible* if for any two different sites

$x,y \in S$ there is a finite sequence $x = x_0, x_1, \ldots, x_n = y$ such that $x_i \neq x_{i-1}$ and

$$\min \{ c(x_{i-1} x_i, \omega) : \omega \in \Omega, \omega_{x_{i-1}} \neq \omega_{x_i} \} > 0$$

for all $1 \leq i \leq n$. For example, the rates (2.6 - 8) are irreducible if there is an irreducible, symmetric matrix $p$ with nonnegative entries such that

$$| \{ a \in F : p_a(x,y) \geq p(x,y) \text{ for all } x,y \in S \} | \geq |F| - 1 .$$

Now we are ready to state the theorem describing the relation between particle jump processes and canonical Gibbs measures.

(2.14) Theorem: *Suppose we are given a rate function* $c(\cdot\,,\cdot)$ *satisfying condition* $(R_\Phi)$ *for some potential* $\Phi$. *Then* $C(\Phi) \subset R(c)$. *If in addition* $c(\cdot\,,\cdot)$ *is irreducible then* $C(\Phi) = R(c)$.

The proof is based on two lemmas which show that both $R(c)$ and $C(\Phi)$ are characterized by integral equations which describe the behaviour of a measure when two sites are transposed.

(2.15) Lemma: *Let* $\mu$ *be a probability measure on* $(\Omega, F)$ *and* $c(\cdot\,,\cdot)$ *a rate function. Then* $\mu \in R(c)$ *iff for all* $xy \subset S$ *and* $f \in C(\Omega)$ *the equation*

$$\int \mu(d\omega)\, c(xy, \omega)\, f({}^{xy}\omega) = \int \mu(d\omega)\, c(xy, \omega)\, f(\omega) \tag{2.16}$$

*holds.*

Proof: 1. Let $\mu \in R(c)$ and $xy \subset S$. It is sufficient to prove (2.16) for functions $f$ of the form $f = 1_{\{X_\Lambda = \zeta\}}$, where $xy \subset \Lambda \in S$ and $\zeta \in \Omega_\Lambda$ with $\zeta_x \neq \zeta_y$. Define $g = 1_{\{X_\Lambda = {}^{xy}\zeta\}}$. We claim that

$$\int f\, G\, g\, d\mu = \int g\, G\, f\, d\mu . \tag{2.17}$$

Indeed, this is an immediate and well-known consequence of the reversibility of $\mu$ and the relation

$$\| Gh - (P_t h - h)/t \| \to 0 \qquad (t \to 0) \quad ,$$

which by the definition of the semigroup is valid for any $h \in C_0(\Omega)$ . For our choice of $f$ and $g$ , however, the equations (2.16) and (2.17) are identical since

$$f\ Gg = f\ c(xy,\cdot\ ) \ , \quad g\ Gf = g\ c(xy,\cdot) \quad .$$

These identities follows from the relations $f(\omega)\ g(\omega) = 0$ , $f(\omega)\ g({}^{xy}\omega) = f(\omega)$ , $f(\omega)\ g({}^{uv}\omega) = 0$ unless $uv = xy$ .

2. Now we assume that (2.16) holds for any $f \in C(\Omega)$ . Then for arbitrary $f,g \in C_0(\Omega)$ we have

$$\int \mu(d\omega)\ c(xy,\ \omega)\ f(\omega)\ g({}^{xy}\omega) = \int \mu(d\omega)\ c(xy,\ \omega)\ f({}^{xy}\omega)\ g(\omega) \quad .$$

Subtracting the integral $\int c(xy,\cdot\ )\ fg\ d\mu$ from both sides of this expression and summing up over all $xy \subset S$ we arrive again at equation (2.17) . This gives that $\mu \in R(c)$ . We sketch the standard argument: By the definition of the closure $\overline{G}$ of the operator $G$ on $C_0(\Omega)$ it follows from (2.17) that also for all $f,g$ in the domain $D(\overline{G})$ of $\overline{G}$ the equation

$$(2.18) \qquad \int f\ \overline{G}g\ d\mu = \int g\ \overline{G}f\ d\mu$$

is true. Then the resolvents $R_\lambda = (\lambda - \overline{G})^{-1}$ $(\lambda > 0)$ are considered. These are bounded operators on $C(\Omega)$ with range in $D(\overline{G})$ . Applying the operator $(\lambda - \overline{G})$ to the functions $R_\lambda f$ and $R_\lambda g$ it is easy to deduce from (2.18) that for all $f,g \in C(\Omega)$ and $n = 1$ the relation

$$\int f\ R_\lambda^n\ g\ \ d\mu = \int g\ R_\lambda^n\ f\ \ d\mu$$

is satisfied. Induction shows that this is true for arbitrary $n$, and the proof is concluded by means of the approximation formula

$$P_t f \;=\; \lim_{\lambda \to \infty} \; e^{-t\lambda} \; \exp [\; t \, \lambda^2 R_\lambda \;] \; f$$

which is part of the proof of the Hille-Yosida theorem. ⌋

The next proposition shows that the relation

$$\mu(\, X_\Lambda = \zeta | E_\Lambda \,) \;=\; \gamma_{\Lambda, N(X_\Lambda)} \; (\zeta | \,\cdot\,) \quad \mu\text{ - f.s.} \qquad (\zeta \in \Omega_\Lambda)$$

is already satisfied for all $\Lambda \in S$ as soon as it is true for sufficiently many two-point sets $\Lambda$ (for example, all sets of diameter $1$ when $S = \mathbb{Z}^d$ ) .

(2.19) <u>Proposition:</u> *Suppose we are given a potential* $\Phi$ *and a probability measure* $\mu$ *on* $(\Omega, F)$ . *Then* $\mu$ *is a canonical Gibbs measure with respect to* $\Phi$ *iff for all* $xy \subset S$ *and* $f \in C(\Omega)$ *the identity*

$$(2.20) \qquad \int \mu(d\omega) \; e^{E(xy,\omega)} \; f(\omega) \;=\; \int \mu(d\omega) \; e^{E(xy,\omega)} \; f({}^{xy}\omega)$$

*holds. Moreover, the relation* $T$ *defined by writing* $(x,y) \in T$ *whenever either* $x = y$ *or for all* $f \in C(\Omega)$ *the equality* (2.20) *is true is transitive.*

<u>Proof:</u> 1. Let $\mu \in C$ and $\Lambda = xy \subset S$ . Obviously, we need to verify (2.20) only for functions $f$ of the form $f = 1_{\{X_\Lambda = \zeta\}} \, g$ , where $\zeta \in \Omega_\Lambda$ and $g$ is a bounded $F_{S \setminus \Lambda}$ - measurable function. Inserting such an $f$ into the l.h.s. of (2.20) we get the expression

$$\int \mu(d\omega) \; g(\omega) \; \gamma_{\Lambda, N(\omega_\Lambda)} \; (\zeta | \omega) \; \exp E_\Lambda \; (\zeta | \omega)$$

$$= \int \mu(d\omega) \; g(\omega) \;\; 1_{\{N(X_\Lambda) = N(\zeta)\}} \;\; (\omega) / Z_{\Lambda, N(\zeta)} \; (\omega)$$

which is invariant under the transposition $\zeta \to {}^{xy}\zeta$ .

2. Conversely, let us assume now that $(x,y) \in T$ for all $xy \subset S$. In order to show that $\mu \in C$ we choose a set $\Lambda \in S$ and a function $f$ of the form $f = 1_{\{X_\Lambda = \zeta\}}\, g$, where $\zeta \in \Omega_\Lambda$ and $g$ is continuous and $E_\Lambda$ - measurable. Then (2.20) yields that for all $\zeta \in \Omega_\Lambda$ and $xy \subset \Lambda$

$$(2.21) \qquad \mu(X_\Lambda = \zeta | E_\Lambda) \exp E(xy, \zeta X_{S \setminus \Lambda})$$

$$= \mu(X_\Lambda = {}^{xy}\zeta | E_\Lambda) \exp E(xy, {}^{xy}\zeta X_{S \setminus \Lambda}) \qquad \mu \text{ - a.s.}$$

is true. We may replace the exponentials here by $E_\Lambda(\zeta | \cdot)$ and $E_\Lambda({}^{xy}\zeta | \cdot)$, respectively, since it is easily verified that for all $\omega$ we have

$$(2.22) \qquad E(xy, \zeta\omega_{S \setminus \Lambda}) - E(xy, {}^{xy}\zeta\omega_{S \setminus \Lambda}) = E_\Lambda(\zeta|\omega) - E_\Lambda({}^{xy}\zeta|\omega) .$$

Therefore we deduce from (2.21) that the function

$$\zeta \to \mu(X_\Lambda = \zeta | E_\Lambda) \exp E_\Lambda(\zeta | \cdot)$$

is a.s. constant on each of the sets $\Omega_{\Lambda,L}$ $(L \in A_\Lambda)$. This proves that for all $\zeta$

$$\mu(X_\Lambda = \zeta | E_\Lambda) = \gamma_{\Lambda, N(X_\Lambda)}(\zeta | \cdot) \qquad \mu \text{ - a.s.}$$

3. Finally, we show that $T$ is transitive. Thus, we assume $x \neq y \neq z \neq x$ and $(x,y) \in T$, $(y,z) \in T$ and ask whether also $(x,z) \in T$. First we observe that $(x,y) \in T$ iff for all $f \in C(\Omega)$

$$\int \mu(d\omega)\, f(\omega) = \int \mu(d\omega)\, e^{E(xy,\omega) - E(xy, {}^{xy}\omega)}\, f({}^{xy}\omega) .$$

Using the relation $(y,z) \in T$ we see that the r.h.s. is equal to the expression

$$\int \mu(d\omega)\, e^{E(yz,\omega) - E(yz, {}^{yz}\omega) + E(xy, {}^{yz}\omega) - E(xy, {}^{(xy)(yz)}\omega)}\, f({}^{(xy)(yz)}\omega) .$$

Now again we use $(x,y) \in T$ and that $^{(xy)(yz)(xy)}\omega = {}^{xz}\omega$ . We then obtain the integral

$$\int \mu(d\omega)\, e^{h(\omega)}\, f({}^{xz}\omega) \quad ,$$

where $$h(\omega) = E(xy, \omega) - E(xy, {}^{xy}\omega) + E(yz, {}^{xy}\omega) -$$

$$- E(yz, {}^{(yz)(xy)}\omega) + E(xy, {}^{(yz)(xy)}\omega) - E(xy, {}^{xz}\omega) .$$

In order to complete the proof that $(x,z) \in T$ we need only show

$$(2.23) \qquad h(\omega) = E(xz, \omega) - E(xz, {}^{xz}\omega) \quad .$$

With a little effort, this can be checked directly from the definition, but it is simpler to argue as follows. Let $\mu \in G(1)$ . Then we know from part 1. of the proof that all three relations $(x,y) \in T$ , $(y,z) \in T$ , $(x,z) \in T$ hold, and the transformations above show that (2.23) is satisfied for $\mu$ - a.e.$\omega$ . But both sides of (2.23) are continuous functions of $\omega$ , and $\mu$ is everywhere dense in $\Omega$ since $\mu$ is positive on all cylinder events. Thus (2.23) holds for all $\omega$ . ⌟

Now we complete the

(2.24) <u>Proof of Theorem (2.14) :</u> Inserting a function f of the form $f = d(xy, \ )g$ into (2.20) we get (2.16) and thus $C(\Phi) \subset R(c)$ . Conversely, (2.16) implies (2.20) whenever $\min \{ d(xy, \omega) : \omega_x \neq \omega_y \} > 0$ . Therefore, if $c(\cdot\, , \cdot)$ is irreducible then from the transitivity of T we get

$$R(c) \subset C(\Phi) \quad . \quad ⌟$$

Finally, we supply the

(2.25) <u>Proof of Remark (2.11) :</u> Suppose $\mu \in R(c) \cap G(z)$ . Then for any two-point set $\Lambda = xy \subset S$ we obtain from (2.16) by particular choices of f that for all $\zeta \in \Omega_\Lambda$ and $\mu$ - a.e.$\omega$

$$\gamma_\Lambda^z(\zeta|\omega)\ c(xy, \zeta\omega_{S\setminus\Lambda}) = \gamma_\Lambda^z({}^{xy}\zeta|\omega)\ c(xy, {}^{xy}\zeta\omega_{S\setminus\Lambda}) \quad .$$

This identity extends to all $\omega$ by continuity since $\mu$ is everywhere dense in $\Omega$. From this we get (2.12) when we cancel the partition function and the activity . ⌋

## 2.2 Particle motions in the continuous model

In this section we will consider two different time evolutions for the continuous model introduced in section 1.2 . These are 1. a diffusion process with interaction, and 2. the Hamilton flow. The canonical Gibbs measures will be shown to be the reversible equilibrium states for the first process and the stable equilibrium states for the second one. But first we will state the continuous analogue of Proposition (2.19) .

In order to do this we need some notation. For $N \geq 0$ and $\omega \in \Omega$, we let $\omega^{(N)}$ denote the measure on $\{N(S) = N\}$ which is defined by

$$(2.26) \qquad \int \omega^{(N)}(d\zeta)\, f(\zeta) = \sum_{\substack{\alpha \subset \omega \\ \alpha(S)=N}} f(\alpha) \quad ,$$

where $f \geq 0$ is an arbitrary measurable function on $\Omega$. Furthermore, we define a measure $\sigma^{(N)}$ on $\{N(S) = N\}$ by the formula

$$(2.27) \qquad \int \sigma^{(N)}(d\eta)\, f(\eta) = \frac{1}{N!} \int \sigma(dx_1) \ldots \int \sigma(dx_N)\, f(\{x_1,\ldots,x_N\}) \quad ,$$

where again $f \geq 0$ is an arbitrary measurable function on $\Omega$. Finally, we extend the notation (1.59) to finite configurations $\zeta$ by putting

$$(2.28) \qquad E(\zeta|\omega) = \sum_{\substack{\alpha \subset \zeta\omega \\ \alpha \cap \zeta \neq \emptyset}} \Phi(\alpha)$$

whenever $\zeta,\eta \in \Omega$ and $\zeta(S) < \infty$ .

(2.29) Proposition: *Given a potential* $\Phi$ *and a probability measure* $\mu$ *on* $(\Omega_{r(\Phi)}, F)$ , *we consider the following three statements:*

(a) $\mu \in C(\Phi)$

(b) *For any* $N \geq 1$ *and any* $F \otimes F \otimes F$ *- measurable function* $F : \Omega \times \Omega \times \Omega \to [0, \infty[$

$$(2.30) \quad \int \mu(d\omega) \int \omega^{(N)}(d\zeta) \int \sigma^{(N)}(d\eta)\, e^{-E(\eta|\omega\setminus\zeta)}\, F(\zeta,\eta,\omega\setminus\zeta)$$

$$= \int \mu(d\omega) \int \omega^{(N)}(d\zeta) \int \sigma^{(N)}(d\eta)\, e^{-E(\eta|\omega\setminus\zeta)}\, F(\eta,\zeta,\omega\setminus\zeta)$$

(c) *For any* $B \otimes B \otimes F$ *- measurable function* $F : S \times S \times \Omega \to [0, \infty[$

$$(2.31) \quad \int \mu(d\omega) \int \omega(dx) \int \sigma(dy)\, e^{-E(y|\omega\setminus x)}\, F(x,y,\omega\setminus x)$$

$$= \int \mu(d\omega) \int \omega(dx) \int \sigma(dy)\, e^{-E(y|\omega\setminus x)}\, F(y,x,\omega\setminus x) \quad .$$

*Then the implications* (a) $\Leftrightarrow$ (b) $\Rightarrow$ (c) *hold. If* $r(\Phi) = 0$ *then all three statements are equivalent.*

The proof of this proposition will be given at the end of the section. It is not difficult to see that (2.30) and (2.31) are satisfied as soon as, for some $z > 0$, $\mu$ obeys the integral equation which, according to a result of Nguyen and Zessin (1976), characterizes the measures in $G(z)$ . Moreover, in the case $\sigma = \lambda$ , $\Phi \equiv 0$ , (2.31) is an immediate consequence of a property which has been introduced by Mecke (1976) in order to characterize the mixed Poisson processes; therefore, Mecke's result follows also from (2.29) and (4.29) below.

* * *

Now we introduce what we have called a diffusion process with interaction. Let $\sigma = \lambda$ , $S = \mathbb{R}^d$ . We choose the set $M_\infty = \{ \omega \in M : \omega(S) = \infty \}$ for the configuration space. By means of polar coordinates, for instance, it is possible to construct a measurable map $a : \omega \to ( a_i(\omega) )_{i\geq 1}$ from $M_\infty$ to $S^{\mathbb{N}}$ such that $\omega = \sum_{i\geq 1} \varepsilon_{a_i(\omega)}$ . This "measurable numbering" will be used to identify each particle during the evolution. We describe the interaction of the particles by a three-times continuously differentiable, symmetric function $\Phi : S \to \mathbb{R}$ such that $\Phi(x) = 0$ if $|x|$ exceeds a certain number $R < \infty$ . Clearly, $\Phi$ defines a shift-invariant pair potential (again denoted by $\Phi$ ) of finite range $R$ by the formula $\Phi(x,y) = \Phi(x - y)$ . The force acting on a particle at $x$ due to the presence of a particle at $y$ is then given by $-\frac{\partial}{\partial x} \Phi(x - y)$ , where $\frac{\partial}{\partial x} \Phi$ denotes the gradient of the function $\Phi$ . Thus, if $\omega \in M_\infty$ and $\omega(\{x\}) \geq 1$ then each particle at $x$ is acted on by the force

$$(2.32) \qquad c_\Phi (x,\omega) = -\int (\omega - \varepsilon_x)(dy) \frac{\partial}{\partial x} \Phi(x - y) \quad .$$

If we imagine that the particles are embedded in a viscous medium of randomly distributed particles then this force determines the mean speed of a particle at $x$ and not its acceleration, and the particles will randomly and independently fluctuate around their mean directions. This type of motion can be described mathematically as follows:

Let $(\beta_i)_{i\geq 1}$ be a sequence of independent $d$ - dimensional standard Wiener processes $(\beta_i(t))_{t\geq 0}$ on a common probability space $(W, \mathcal{W}, P)$ , $\mu$ a probability measure on $(M_\infty, F)$ , and $(\xi(t))_{t\geq 0}$ a Markov process with state space $M_\infty$ and initial distribution $\mu$ which is defined on $( M_\infty \times W, F \otimes \mathcal{W}, \mu \otimes P )$ . $(\xi(t))_{t\geq 0}$ will be called a *diffusion process with interaction* $\Phi$ if for $\mu \otimes P$ - a.e. $(\omega,w)$ the function $t \to \xi( t, \omega, w )$ can be written in the form

$$(2.33) \qquad \xi( t, \omega, w ) = \sum_{i\geq 1} \varepsilon_{x_i(t,\omega,w)} \qquad (t \geq 0)$$

where $( x_i(\cdot , \omega, w) )_{i\geq 1}$ is a solution of the system of equations

$$(2.34) \qquad x_i(t, \omega, w) - a_i(\omega) = \frac{1}{2} \int_0^t ds \; c_\Phi( x_i(s, \omega, w), \xi(s, \omega, w) ) + \beta_i(t, w) \qquad ( t \geq 0, \; i \geq 1).$$

Such a process is said to be *reversible* if for all $A, B \in F$ and $t > 0$

$$\mu \otimes P \; ( \xi(0) \in A, \; \xi(t) \in B ) = \mu \otimes P \; ( \xi(0) \in B, \; \xi(t) \in A ) \quad .$$

The following theorem on the existence and characterization of reversible diffusion processes with interaction is due to R. Lang (1977). Instead of reproducing his precise technical conditions we will only describe their type and refer to his paper for details. The terms "superstable" and "tempered" are used in the sense of Ruelle (1970).

(2.35) Theorem: (i) *Suppose that* $\Phi$ *is superstable,* $z > 0$ , *and* $\mu \in G(z, \Phi)$ *is tempered. Then there exists an (essentially unique) diffusion process with interaction* $\Phi$ *and initial distribution* $\mu$ . *This process is reversible.*

(ii) *Suppose that* $\mu$ *is a probability measure on* $(M_\infty, F)$ *which locally has a sufficiently smooth Radon-Nikodym density with respect to* $\pi$ . *Moreover, suppose that the particle density and the logarithmically normalized particle fluctuation are functions in* $\bigcap_{p\geq 1} L^p(\mu)$ . *Assume that there is a reversible diffusion process with interaction* $\Phi$ *and initial distribution* $\mu$ . *Then* $\mu \in C(\Phi)$ .

It is not difficult to see that for shift-invariant pair potentials, as considered here, condition (D) of § 6 is satisfied. According to Theorem (6.14) this implies that any $\mu \in C(\Phi)$ is a mixture of Gibbs measures in $\bigcup_z G(z, \Phi)$ . Therefore Theorem (2.35) can be restated as follows: Suppose that $\Phi$ is a smooth, shift-invariant, superstable pair potential of finite range and $\mu$ is a sufficiently reasonable probability measure on $\Omega_0$ . Then $\mu \in C(\Phi)$ iff there is a reversible diffusion process with respect to $\Phi$ and $\mu$ .

We are not going to give a complete proof of this theorem. Rather we will prove a proposition which, apart from the existence result, contains the essence of Theorem (2.35). Notice that part (i) of (2.35) does not ensure the existence of a Markov semigroup of transition kernels. But if such a semigroup existed then its generator $G$ would certainly be given by

$$(2.36) \qquad 2\, Gf(\omega) = \int \omega(dx)\, [\, (\tfrac{\partial}{\partial x})^2 f(\omega) + c_\Phi(x,\omega) \tfrac{\partial}{\partial x} f(\omega)\, ]$$

whenever this expression is well-defined. Clearly this is the case when $f$ belongs to $C^2_{loc}(M)$. Here for any $k \in \{1, 2, \dots, \infty\}$ we let $C^k_{loc}(M)$ denote the set of all real functions $f$ on $M$ with the properties: For some $\Lambda \in S$, $f$ is $F_\Lambda$-measurable, and for every $n \geq 1$ the function $(x_1,\dots,x_n) \to f(\varepsilon_{x_1}+\dots+\varepsilon_{x_n})$ on $S^n$ is $k$ times continuously differentiable; if $\omega(\{x\}) \geq 1$ then $\frac{\partial}{\partial x} f(\omega)$ denotes the gradient of the function $y \to f(\omega + \varepsilon_y - \varepsilon_x)$ at $y = x$. Furthermore, a diffusion process with interaction $\Phi$ and initial distribution $\mu$ would be reversible iff there was a sufficiently large class $\underline{\underline{A}} \subset C^2_{loc}(M)$ with the property:

(2.37) If $f,g \in \underline{\underline{A}}$ and $f\, Gg \in L^1(\mu)$ then $g\, Gf \in L^1(\mu)$ and $\int f\, Gg\, d\mu = \int g\, Gf\, d\mu$.

R. Lang has shown that for a certain class $\underline{\underline{A}}$ the implication "only if" actually holds. What we will prove here is the fact that the detailed balance equation (2.37) is equivalent to the statement (2.29) (c) and therefore to the relation $\mu \in C$. This result is not restricted to shift-invariant pair potentials. A general potential $\Phi$ of finite range and particle diameter $r \geq 0$ is said to be continuously differentiable if for every $n \geq 1$ the mapping $(x_1,\dots,x_n) \to \Phi(\{x_1,\dots,x_n\})$ on

$$S^n_r = \{ (x_1,\dots,x_n) \in S^n : |x_i - x_j| > r \ (i \neq j) \}$$

is continuously differentiable. For such a $\Phi$ and $f \in C^1_{loc}(\Omega)$, $Gf$ is defined on $\Omega$ by means of the function

$$c_\Phi(x,\omega) = -\frac{\partial}{\partial x} E(x|\omega \setminus x) \qquad (x \in \omega).$$

Here $C^k_{loc}(\Omega) = \{ f|\Omega : f \in C^k_{loc}(M) \}$ .

(2.38) Proposition: *Suppose that* $\Phi$ *is a continuously differentiable potential of finite range such that* $r(\Phi) = 0$ *and* $\mu$ *is a probability measure on* $(\Omega_0, F)$ .

(i) *If* $\mu \in C(\Phi)$ *then* (2.37) *holds for the algebra* $\underline{\underline{A}} = C^2_{loc}(\Omega)$ .

(ii) *Let* $\underline{\underline{A}}$ *denote the algebra generated by all functions of the form* $f(\omega) = \psi( \int \varphi \, d\omega )$ *with* $\varphi$ *a real* $C^\infty$ *- function on* $S$ *having compact support and* $\psi$ *a bounded* $C^\infty$ *- function on* $\mathbb{R}$ . *Assume that* $fGg \in L^1(\mu)$ *whenever* $f,g \in \underline{\underline{A}}$ . *Furthermore, suppose that for every* $\Lambda \in S$ $\mu$ *is absolutely continuous with respect to* $\pi$ *on* $F_\Lambda$ *with density* $u_\Lambda$ . *Assume* $u_\Lambda \in C^1_{loc}(\Omega)$ *and*

$$(2.39) \qquad \frac{\partial}{\partial x} u_\Lambda(x\omega) = \int \pi(d\zeta) \frac{\partial}{\partial x} u_\Delta( x\omega_\Lambda \, \zeta_{\Delta\setminus\Lambda} ) \qquad (\omega \in \Omega)$$

*whenever* $x \in V$ *and* $V \subset \Lambda \subset \Delta \in S$ *are such that* $\Phi(\alpha) = 0$ *unless* $\alpha(V)\,\alpha(\Delta \setminus \Lambda) = 0$ . *Then* (2.37) *implies that* $\mu \in C(\Phi)$ .

Notice that (2.39) is a condition about the change of order of differentiation and integration, since

$$(2.40) \qquad u_\Lambda(\omega) = \int \pi(d\zeta) \; u_\Delta(\omega_\Lambda \, \zeta_{\Delta\setminus\Lambda}) \qquad (\Lambda \subset \Delta)$$

for $\pi$ - a.e.$\omega$ and therefore, by continuity, for all $\omega$ . It is easily seen that (2.39) is satisfied when $\mu \in C(\Phi)$ .

Proof: "(i)" : Choose functions $f,g \in C^2_{loc}(\Omega)$ with $fGg \in L^1(\mu)$ and a set $\Lambda \in S$ such that $f$ and $g$ are $F_\Lambda$ - measurable. Define a measurable function $F$ on $S \times S \times \Omega$ by

$$F( x, y, \omega ) = e^{E(y|\omega)} \, 1_\Lambda(y) \, f(x\omega) \, [ \frac{\partial^2}{\partial x^2} g(x\omega) + c_\Phi(x,\omega) \frac{\partial}{\partial x} g(x\omega) ] .$$

Then

$$2 \;\; \lambda(\Lambda) \int f \; Gg \; d\mu \;\; =$$

$$= \int \mu(d\omega) \int \omega(dx) \int \lambda(dy) \;\; e^{-E(y|\omega \setminus x)} \;\; F(x, y, \omega \setminus x) \qquad .$$

In particular, the latter integral exists. Therefore (2.31) applies showing that this integral coincides with

$$\int \mu(d\omega) \int_{\Lambda} \omega(dx) \;\; e^{E(x|\omega \setminus x)} \int \lambda(dy) \; f(y\omega \setminus x) \; \frac{\partial}{\partial y} \left[ \frac{\partial}{\partial y} g(y\omega \setminus x) \; e^{-E(y|\omega \setminus x)} \right] \; .$$

By partial integration this is seen to be equal to

$$- \int \mu(d\omega) \int_{\Lambda} \omega(dx) \;\; e^{E(x|\omega \setminus x)} \int \lambda(dy) \;\; e^{-E(y|\omega \setminus x)} \; \frac{\partial}{\partial y} f(y\omega \setminus x) \; \frac{\partial}{\partial y} g(y\omega \setminus x)$$

which is symmetric in $f$ and $g$ .

"(ii)" We choose a closed cube $V$ and an open cube $\Lambda \supset V$ such that $\Phi(\alpha) = 0$ whenever $\alpha(V)\,\alpha(S \setminus \Lambda) > 0$ . Let $\varepsilon > 0$ and $0 \leq \varphi_1 \leq 1$ be a $C^\infty$ - function on $S$ which equals $1$ on $\partial_\varepsilon V = \{ x \in V : |x - y| \leq \varepsilon$ for some $y \notin V \}$ and $0$ on $\{ y \in S : |y - x| \geq \delta$ for all $x \in \partial_\varepsilon V \}$ for some $\delta > 0$ . Let $0 \leq \varphi_1 \leq 1$ be a $C^\infty$ - function on $\mathbb{R}$ with $\varphi_1(0) = 1$ and $\varphi_1(s) = 0$ $(s \geq 1)$ . Then $f_1(\omega) = \varphi_1( \int \varphi_1 \, d\omega )$ is a "smooth indicator function" of the event $\{N(\partial_\varepsilon V) = 0\}$. Next we choose a number $N \geq 1$ , a smooth approximation $\varphi_2$ of $1_\Lambda$ and a smooth approximation $\varphi_2$ of $1_{[0,N]}$ such that $f_2(\omega) = \varphi_2( \int \varphi_2 \, d\omega )$ is a "smooth indicator function" of the event $\{ N(\Lambda) \leq N \}$ . Moreover, we let $\varphi_3 \geq 0$ be a $C^\infty$ - function on $S$ with support in $\Lambda$ , and put $f_3(\omega) = \exp[ - \int \varphi_3 \, d\omega ]$ . We can (and will) assume that $\varphi_2 = 1$ on the support of $\varphi_3$ . Finally, we choose a coordinate axis $1 \leq i \leq d$ , a closed interval $I \subset \mathbb{R}$ , a smooth approximation $0 \leq \overline{\varphi}_4 \leq 1$ of $1_I$ , and a $C^\infty$ - function $\varphi_4$ on $S$ with support in $V$ such that $\varphi_4(x) = \int_{-\infty}^{x^i} \overline{\varphi}_4(s)\, ds$ whenever $x = ( x^1, \ldots, x^d ) \in V \setminus \partial_\varepsilon V$ . We put $f_4(\omega) =$

$= \psi_4( \int \varphi_4 \, d\omega )$ , where $\psi_4$ is a bounded $C^\infty$ - function on $\mathbb{R}$ such that $\psi_4(s) = s$ whenever $|s| \leq N \max_{x \in V} |x^i|$ .

Now let

$$f(\omega) = \sqrt{f_1(\omega) \, f_2(\omega) \, f_3(\omega)}$$

$$g(\omega) = f(\omega) \, f_4(\omega) .$$

Then $f,g \in \underline{\underline{A}}$ , and for sufficiently large $\Delta \supset \Lambda$ we have

$$\int fGg \, d\mu =$$

$$= \int \lambda(dx) \int \pi(d\omega) \; u_\Delta(x\omega) \; f(x\omega) \; e^{E(x|\omega)} \frac{\partial}{\partial x} [ \frac{\partial}{\partial x} g(x\omega) \, e^{-E(x|\omega)} ] .$$

Here we have used that $\mu = u_\Delta \pi$ on $F_\Delta$ , and (1.57) . By partial integration we get for the last expression

$$- \int \lambda(dx) \int \pi(d\omega) \, e^{-E(x|\omega)} \frac{\partial}{\partial x} g(x\omega) \frac{\partial}{\partial x} [ f(x\omega) \, u_\Delta(x\omega) \, e^{E(x|\omega)} ]$$

$$= - \int \lambda(dx) \int \pi(d\omega) \; u_\Delta(x\omega) \frac{\partial}{\partial x} g(x\omega) \frac{\partial}{\partial x} f(x\omega)$$

$$- \int \lambda(dx) \int \pi(d\omega) \, e^{-E(x|\omega)} \frac{\partial}{\partial x} g(x\omega) \, f(x\omega) \frac{\partial}{\partial x} [ u_\Delta(x\omega) \, e^{E(x|\omega)} ]$$

$$= I + II .$$

(Notice that the integral I and therefore also II exists since the function $f_2$ is incorporated in $f$ and $g$ .) By assumption I + II remains unchanged if $f$ and $g$ are interchanged. But I is symmetric in $f$ and $g$ . Hence so is II .

Now observe that

$$\frac{\partial}{\partial x} g(x\omega) = f_4(x\omega) \frac{\partial}{\partial x} f(x\omega) + f(x\omega) \frac{\partial}{\partial x} f_4(x\omega) .$$

Inserting this sum into II we get from the symmetry of II , (1.57) , and the definition of $f_2$ and $f_4$ .

$$\int \pi(d\omega)\ f_1(\omega)\ f_2(\omega)\ f_3(\omega) \int_V \omega(dx)\ e^{-E(x|\omega\setminus x)}\ \overline{\varphi}_4(x^i)\ \frac{\partial}{\partial x^i} [\ u_\Delta(\omega)\ e^{E(x|\omega\setminus x)}\ ]$$

$$= \quad 0$$

Now we take the limits $\varphi_1 \downarrow 1_{\partial_\varepsilon V}$ , $\varphi_2 \uparrow 1_\Lambda$ , $\overline{\varphi}_4 \downarrow 1_I$ , and $\varepsilon \downarrow 0$ . Then by dominated convergence we have

$$\int_{\{N(\Lambda)\leq N\}} \pi(d\omega)\ f_3(\omega) \int_V \omega(dx)\ e^{-E(x|\omega\setminus x)}\ 1_I(x^i)\ \frac{\partial}{\partial x^i} [\ u_\Delta(\omega)\ e^{E(x|\omega\setminus x)}\ ] = 0$$

and (1.56) , (2.39) , and (2.40) give

$$\int_{\{N(\Lambda)\leq N\}} \pi(d\omega)\ f_3(\omega) \int_V \omega(dx)\ e^{-E(x|\omega\setminus x)}\ 1_I(x^i)\ \frac{\partial}{\partial x^i} [\ u_\Lambda(\omega)\ e^{E(x|\omega\setminus x)}\ ] = 0 .$$

As a function of $f_3$ the last integral is a signed measure on $F_\Lambda$ , and we have shown that its Laplace transform vanishes. Since $N$ is arbitrary this leads to the conclusion that for $\pi$ - a.e.$\omega$ and all rational intervals $I$

$$\int_V \omega(dx)\ e^{-E(x|\omega\setminus x)}\ 1_I(x^i)\ \frac{\partial}{\partial x^i} [\ u_\Lambda(\omega)\ e^{E(x|\omega\setminus x)}\ ] = 0 .$$

Now observe that $\pi$ - a.s. the $i$ - th coordinates of the particles in $V$ are distinct. Thus for $\pi$ - a.e.$\omega$ and every $x \in \omega_V$ we have

$$\frac{\partial}{\partial x^i} [\ u_\Lambda(\omega)\ e^{E(x|\omega\setminus x)}\ ] = 0 .$$

By continuity we see that for all $\omega$ and all $x \in V \setminus \omega$

$$\frac{\partial}{\partial x^i} [\ u_\Lambda(x\omega)\ e^{E(x|\omega)}\ ] = 0 .$$

Since $i$ was arbitrary we have the result: For all $\omega \in \Omega$ there is a constant $c(\omega)$ such that

$$u_\Lambda(x\omega) = c(\omega)\; e^{-E(x|\omega)}$$

whenever $x \in V \setminus \omega$. This implies (2.31). Indeed, let $F(x, y, \omega) \geq 0$ be a $\mathcal{B} \otimes \mathcal{B} \otimes \mathcal{F}_\Lambda$ - measurable function with $F(x, y, \omega) = 0$ whenever $x \notin V$ or $y \notin V$. Then

$$\int \mu(d\omega) \int \omega(dx) \int \lambda(dy)\; e^{-E(y|\omega \setminus x)}\; F(x, y, \omega \setminus x)$$

$$= \int \pi(d\omega)\; c(\omega) \int \lambda(dx)\; e^{-E(x|\omega)} \int \lambda(dy)\; e^{-E(y|\omega)}\; F(x, y, \omega) \quad .$$

This expression is symmetric in $x$ and $y$, and since $V$ and $\Lambda$ are arbitrary we conclude that (2.29) (c) holds. Hence $\mu \in \mathcal{C}(\Phi)$. ⌋

* * *

Now we consider the classical time evolution of a particle system, namely the one determined by Hamilton's equations. In these equations a particle is characterized not only by its position $q \in \mathbb{R}^d$ but also by its momentum $p \in \mathbb{R}^d$. Therefore we define $S = \mathbb{R}^d \times \mathbb{R}^d$ and think of each $x \in S$ as a pair $(q_x, p_x)$ of position and momentum coordinates. Furthermore, we choose $\sigma = \lambda \otimes \nu$, where $\lambda$ is the $d$ - dimensional Lebesgue measure and $\nu$ denotes the $d$ - dimensional standard normal distribution. In order to adapt the definitions of section 1.2 to this new interpretation of $S$ we introduce the semi-norm $\| x \| = |q_x|$ on $S$ and the system $\mathcal{S}_q$ of all sets of the form $\Lambda = V \times \mathbb{R}^d$, where $V \subset \mathbb{R}^d$ is a bounded Borel set such that $\lambda(V) > 0$. Moreover, we will say that a potential is standard if

(2.41) $$\Phi(\alpha) = \Phi(\{(q_x, 0) : x \in \alpha\}) \qquad (\alpha \in \Omega_f) \quad .$$

If the seminorm $\| \cdot \|$ is used to describe the shape of the particles then the definitions of the spaces $\Omega_{(r)}$ and $\Omega_r$ immediately carry over to the present context, and, once $S$ is replaced by $S_q$ and only standard potentials are considered, then so do the definitions of the Poisson process, of potentials and of canonical and grand canonical Gibbs distributions and measures.

From now on we fix a continuously differentiable standard potential $\Phi$ of finite range and particle diameter $r \geq 0$ . (If $r > 0$ then the finite range condition is not necessary; what we really need is the absolute convergence of

$$\frac{\partial}{\partial x} E( x | \omega \setminus x ) = \sum_{x \in \alpha \subset \omega} \frac{\partial}{\partial x} \Phi(\alpha)$$

whenever $x \in \omega \in \Omega_r$ ) .

For any two functions $f, g \in C^1_{loc}(\Omega)$ the *Poisson bracket* is defined by

$$(2.42) \qquad \{f,g\} (\omega) = \sum_{x \in \omega} [ \frac{\partial f}{\partial q_x} (\omega) \frac{\partial g}{\partial p_x} (\omega) - \frac{\partial f}{\partial p_x} (\omega) \frac{\partial g}{\partial q_x} (\omega) ] .$$

Here $\frac{\partial f}{\partial q_x}(\omega)$ denotes the gradient of the function $q \to f( \{(q, p_x)\} \cup \omega \setminus \{x\} )$ at $q = q_x$ ; $\frac{\partial f}{\partial p_x}(\omega)$ is similarly defined. Furthermore, by means of the formal Hamiltonian

$$(2.43) \qquad H(\omega) = \sum_{x \in \omega} |p_x|^2 / 2 + \sum_{\alpha \subset \omega} \Phi(\alpha)$$

(the particle mass is normalized to be 1 )

we introduce the Poisson bracket

$$(2.44) \qquad \{f,H\} (\omega) =$$

$$= \sum_{x \in \omega} [ \frac{\partial f}{\partial q_x} (\omega) p_x - \frac{\partial f}{\partial p_x} (\omega) \sum_{x \in \alpha \subset \omega} \frac{\partial \Phi}{\partial q_x} (\alpha) ]$$

which is well-defined if $f \in C^1_{loc}(\Omega)$ .

The physical significance of the bracket (2.44) lies in the fact that it defines the generator of the Hamilton flow whenever this exists. More precisely, the following is true. For certain reasonable shift-invariant pair potentials $\Phi$ it is known (see, for instance, Fritz/Dobrushin (1977), Lanford (1975), Presutti et al. (1976) ) that there is a measurable subset $\hat{M}$ of $M$ with the properties:

(i) $\mu(\hat{M}) = 1$ for all $\mu \in \bigcup_{z>0} G(\Phi, z)$ .

(ii) For every $\omega \in \hat{M}$ , the *Hamilton equations*

$$(2.45) \qquad \dot{q}_i = p_i \ , \quad \dot{p}_i = - \sum_{j \neq i} \frac{\partial}{\partial q_i} \Phi( q_i - q_j ) \qquad (i \geq 1)$$

have an (essentially unique) solution $( q_i(t, \omega) , p_i(t, \omega) )_{i \geq 1, t \in \mathbb{R}}$ with initial condition $( q_i(0, \omega) , p_i(0, \omega) ) = a_i(\omega)$ $(i \geq 1)$ .

(iii) The mapping

$$(t, \omega) \rightarrow T_t\omega = \sum_{i \geq 1} \varepsilon_{(q_i(t,\omega), p_i(t,\omega))}$$

defines a flow on $\hat{M}$ .

(iv) Every $\mu \in \bigcup_z G(\Phi, z)$ is invariant with respect to the flow $(T_t)_{t \in \mathbb{R}}$ .

Moreover, if $\Phi$ has finite range or a hard core then a slight modification of the results of § 6 shows that every $\mu \in C(\Phi)$ is a mixture of Gibbs measures and therefore the properties (i) and (iv) extend to all $\mu \in C(\Phi)$ . For these $\mu$ the flow $(T_t)_{t \in \mathbb{R}}$ induces a strongly continuous group $(U_t)_{t \in \mathbb{R}}$ of unitary operators acting on the complex Hilbert space $L^2(\mu)$ . Thus Stone's theorem (see Reed/Simon (1972), for instance) implies that there is a self-adjoint operator $G$ such that $U_t = e^{itG}$ and

$$(2.46) \qquad i\,G\,f = \lim_{t \to 0} (U_t f - f) / t \qquad (f \in D(G)) \ .$$

Now it is easily seen from (2.45) that every bounded $f \in C^1_{loc}(\Omega)$ belongs to the domain $D(G)$ of $G$ and satisfies

$$(2.47) \qquad i\,G\,f \;=\; \{\, f,\, H \,\} \quad .$$

This explains the dynamical significance of the bracket $\{f, H\}$ . In the following we will use this bracket in order to give a second dynamical characterization of canonical Gibbs measures.

(2.48) Remark: *Suppose that* $\Phi$ *is a continuously differentiable standard potential of finite range and particle diameter* $r \geq 0$ . *If* $r > 0$ *then in addition assume that* $\Phi(\{x, y\}) \to \infty$ *whenever* $\|x - y\| \to r$ . *Let* $\mu \in C(\Phi)$ . *Then*

$$(2.49) \qquad \int \{f, g\}\, d\mu \;=\; \int g\, \{f, H\}\, d\mu$$

*provided* $f,g \in C^2_{loc}(\Omega)$ *are such that* $\{f, g\}$ *and* $g\{f, H\}$ *are integrable .*

Proof: Let $H(x|\omega) = |p_x|^2/2 + E(x|\omega)$ and

$$h(x, \omega) \;=\; \frac{\partial f}{\partial q_x}(x\omega)\, \frac{\partial}{\partial p_x}\left(g(x\omega)\, e^{-H(x|\omega)}\right) \;-\; \frac{\partial f}{\partial p_x}(x\omega)\, \frac{\partial}{\partial p_x}\left(g(x\omega)\, e^{-H(x|\omega)}\right)$$

whenever $x \notin \omega$ . Then

$$\int \{f, g\}\, d\mu \;-\; \int g\, \{f, H\}\, d\mu \;=$$

$$= \int \mu(d\omega) \int \omega(dx)\, e^{H(x|\omega \setminus x)}\; h(\, x,\, \omega \setminus x\,) \quad .$$

Now we choose some $\Lambda \in S$ and introduce the function

$$F(\, x,\, y,\, \omega\,) \;=\; e^{H(x|\omega)}\; e^{E(y|\omega)}\; 1_\Lambda(y)\; 1_{\Omega_{(r)}}(x\omega)\; 1_{\Omega_{(r)}}(y\omega)\; h(x, \omega) \quad .$$

Then the last integral is seen to be a positive multiple of

$$\int \mu(d\omega) \int \omega(dx) \int \sigma(dy)\ e^{-E(y|\omega\setminus x)}\ F(x, y, \omega \setminus x) \ .$$

According to (2.29) (c) we may interchange the arguments $x$ and $y$ in $F$. This shows that the difference between the two sides of (2.49) is a non-zero multiple of

$$\int \mu(d\omega) \int_\Lambda \omega(dx)\ e^{E(x|\omega\setminus x)} \int \lambda \otimes \lambda(dy)\ 1_{\Omega_r}(y\omega \setminus x)\ \ h(y, \omega \setminus x) \ .$$

Thus we need only to show that for all $\omega$ and $x \in \omega$ the inner integral vanishes. This is done using integration by parts: Solve for the function $y \to g_0(y) = g(y\,\omega \setminus x)\ e^{-H(y|\omega\setminus x)}$ and observe that (i) for sufficiently large $|y|$ all partial derivatives of $y \to f(y\,\omega \setminus x)$ vanish, and (ii) if $r > 0$ then $g_0(y)$ vanishes on the boundaries of the holes which are enclosed in the domain of integration. ┘

There are three reasons for the interest in equation (2.49). First, we will show below that canonical Gibbs states are actually characterized by this equation. Secondly, (2.49) is a classical analogue of the Kubo-Martin-Schwinger condition which is well-known in quantum mechanics. This is the origin of the name given in the following

(2.50) <u>Definition:</u> Let $\underline{\underline{A}} \subset C^1_{loc}(\Omega)$ be an algebra. A probability measure $\mu$ on $(\Omega_r, F)$ is called a (static) KMS - *state* with respect to $\underline{\underline{A}}$ and $\Phi$ if for any $f,g \in \underline{\underline{A}}$ the functions $\{f, g\}$ and $g\{f, H\}$ are integrable and

$$\int \{f, g\}\ d\mu \quad = \quad \int g\{f, H\}\ d\mu \ .$$

The third reason is the following result demonstrating the dynamical significance of the KMS condition; it was proved by Aizenman et al. (1976), Pulvirenti/Riela (1977), and Pulvirenti (1977) under different technical assumptions (one of which was verified for particular situations by Marchioro et al. (1978) ) : Suppose that

$\mu$ is a probability measure on $(\Omega, F)$ for which the Hamilton flow with respect to $\Phi$ a.s. exists. Assume that $\mu$ is invariant under this flow and, in a certain strong sense, stable under perturbations of the Hamiltonian H by functions in a suitable algebra $\underline{\underline{A}} \subset C^1_{loc}(\Omega)$ . Moreover, suppose that the expected kinetic energy per particle is 1 . Then $\mu$ is a KMS - state with respect to $\underline{\underline{A}}$ and $\Phi$ .

Now we will state the converse of Remark (2.48) . To this end we need to specify an algebra $\underline{\underline{A}} \subset C^1_{loc}(\Omega)$ which should be as small as possible. A natural choice would be the algebra $\underline{\underline{A}}$ of all functions f of the form

(2.51) $$f(\omega) = \tilde{\varphi}(\omega) = \sum_{\zeta \subset \omega} \varphi(\zeta) = \sum_{n \geq 0} \int \omega^{(n)}(d\zeta)\ \varphi(\zeta) \quad ,$$

where $\varphi$ is a function on $(\Omega_f \cap \Omega_r) \cup \{\emptyset\}$ such that

(2.51 a) for all $n \geq 1$ , the mapping $(x_1,\ldots,x_n) \rightarrow \varphi(\{x_1,\ldots,x_n\})$ is a $C^\infty$ - function on $S^n_r$ with compact support, and

(2.51 b) for all but finitely many n , $\varphi = 0$ on $\{ N(S) = n \}$ .

Unfortunately, $\underline{\underline{A}}$ is a bit too small. Thus we choose the algebra $\bar{\underline{\underline{A}}}$ generated by $\underline{\underline{A}}$ and all functions of the form

(2.52) $$f(\omega) = \exp [ - \int \varphi \, d\omega ] \quad ,$$

where $\varphi$ is a nonnegative $C^\infty$ - function on S with compact support. Actually, $\bar{\underline{\underline{A}}}$ is only a slight enlargement of $\underline{\underline{A}}$ since every function f of the form (2.52) can be written as

$$f(\omega) = \sum_{\zeta \subset \omega} \prod_{x \in \zeta} ( e^{-\varphi(x)} - 1 ) \quad ,$$

which is of the form (2.51) except for the requirement (2.51 b) . Furthermore, it is not difficult to verify that $\bar{\underline{\underline{A}}}$ (just as $\underline{\underline{A}}$ ) has the desirable property that $\{f, g\} \in \bar{\underline{\underline{A}}}$ whenever $f,g \in \bar{\underline{\underline{A}}}$ .

Finally, if $r > 0$ then we will say that a probability measure $\mu$ on $(\Omega_r, F)$ is *loosely packed* if for every connected $\Lambda \in S_q$ and $\mu$ - a.e. $\omega \in \Omega_r$ there is a connected set $\Delta \in S_q$ which contains $\Lambda$ and has the following property: Whenever $\zeta, \eta \subset \Lambda$ are two configurations such that $\zeta(\Lambda) = \eta(\Lambda) = \omega(\Lambda)$ and $\zeta\, \omega_{S \setminus \Lambda} \in \Omega_{(r)}$, $\eta\, \omega_{S \setminus \Lambda} \in \Omega_{(r)}$ then $\zeta\, \omega_{\Delta \setminus \Lambda}$ and $\eta\, \omega_{\Delta \setminus \Lambda}$ belong to the same connected component of

$$\{ \alpha \subset \Delta \ : \ \alpha(\Delta) = \zeta(\Lambda) + \omega(\Delta \setminus \Lambda),\ \alpha\, \omega_{S \setminus \Delta} \in \Omega_{(r)} \} \quad .$$

(2.53) <u>Theorem:</u> *Suppose that* $\Phi$ *is a continuously differentiable standard potential of finite range and with particle diameter* $r \geq 0$ . *Let* $\mu$ *be a* KMS - *state with respect to* $\underline{\underline{A}}$ *and* $\Phi$ . *In the case* $r > 0$ *we assume* $\mu$ *to be loosely packed. Then* $\mu \in C(\Phi)$ .

It is worth noting that it is not necessary to assume $\mu$ is locally absolutely continuous with respect to $\pi$ .

<u>Proof:</u> 1. The *first step* in the proof is to show that $\mu$ satisfies a local KMS - condition. To this end we fix a connected open set $\Lambda \in S_q$ and choose two functions $\varphi$ and $\psi$ on $(\Omega_f \cap \Omega_r) \cup \{\emptyset\}$ satisfying (2.51 a,b) . We assume that the S - support of $\varphi$ and $\psi$ is contained in $\Lambda$ , the S - support of $\varphi$ being defined as the smallest compact set $K \subset S$ such that $\varphi(\zeta) = 0$ unless $\zeta(S \setminus K) = 0$. Furthermore, we let $\chi$ denote a nonnegative $C^\infty$ - function on S which has a compact support and is constant on the S - supports of $\varphi$ and $\psi$ , and put $f(\omega) = \exp[-\int \chi \, d\omega]$ . Then $\{ \tilde{\varphi}, f \} = 0$ and hence

$$\{ \tilde{\varphi}, \tilde{\psi} f \} = \{ \tilde{\varphi}, \tilde{\psi} \} f \quad .$$

Thus we obtain from the KMS - condition, applied to the functions $\tilde{\varphi} \in \underline{\underline{A}}$ , $\tilde{\psi} f \in \underline{\underline{A}}$ ,

$$\int f [ \{ \tilde{\varphi}, \tilde{\psi} \} - \tilde{\psi} \{ \tilde{\varphi}, H \} ] \, d\mu = 0 \quad . \tag{2.54}$$

Next we choose arbitrary compact sets $\Delta_1,\dots,\Delta_n \subset S \setminus \Lambda$ and numbers $a_0,\dots,a_n \geq 0$ and let $\chi$ tend to the step function

$$a_0 \; 1_\Lambda \; + \; \sum_{i=1}^{n} a_i \; 1_{\Delta_i} \quad .$$

By the dominated convergence theorem (which applies since by assumption $\{\, \widetilde{\varphi}, \widetilde{\psi} \,\}$ and $\widetilde{\psi}\,\{\, \widetilde{\varphi}, H \,\}$ are integrable) we get from (2.54) that the Laplace transform of the signed measure

$$( k_0,\dots,k_n ) \rightarrow \int_{\{\, N(\Lambda)=k_0, N(\Delta_1)=k_1,\dots,N(\Delta_n)=k_n \,\}} [\, \{\, \widetilde{\varphi}, \widetilde{\psi} \,\} \; - \; \widetilde{\psi}\,\{\, \widetilde{\varphi}, H \,\} \,] \; d\mu$$

vanishes. This in turn implies that for all sets $A$ in a generator of $E_\Lambda$, and hence for all $A \in E_\Lambda$,

$$\int_A [\, \{\, \widetilde{\varphi}, \widetilde{\psi} \,\} \; - \; \widetilde{\psi}\,\{\, \widetilde{\varphi}, H \,\} \,] \; d\mu \; = \; 0 \quad .$$

This leads to the conclusion that for $\mu$ - a. e. $\omega$

$$(2.55) \qquad \int [\, \{\, \widetilde{\varphi}, \widetilde{\psi} \,\} \; - \; \widetilde{\psi}\,\{\, \widetilde{\varphi}, H_{\Lambda,\omega} \,\} \,] \; d\mu \, ( \, . \mid E_\Lambda \, ) \; (\omega) \; = \; 0 \quad .$$

Here

$$H_{\Lambda,\omega}\,(\zeta) \; = \; \sum_{x \in \zeta_\Lambda} |p_x|^2 \,/\, 2 \; + \; E_\Lambda(\zeta|\omega) \quad ,$$

and we have used the fact that

$$\{\, \widetilde{\varphi}, H \,\} \, ( \, \zeta_\Lambda \, \omega_{S\setminus\Lambda} \, ) \; = \; \{\, \widetilde{\varphi}, H_{\Lambda,\omega} \,\} \; (\zeta) \quad .$$

By a separability and density argument we see that for $\mu$ - a. e. $\omega$ (2.55) is satisfied simultaneously for all those $\varphi$ and $\psi$ whose $S$ - supports are contained in $\Lambda$ and which are only of class $C^1$ instead of $C^\infty$ (Note that uniform convergence of $\varphi$ and its derivatives entails uniform convergence of $\tilde{\varphi}$ and its derivatives on sets of the form $\{ N(\Lambda) = \omega(\Lambda) \}$. )

2. The *main part* of the proof is to deduce from (2.55) that there is an $F_\Lambda \otimes E_\Lambda$-measurable function $\rho_\Lambda( \cdot \mid \cdot )$ such that for $\mu$ - a. e. $\omega$

(2.56 a) $$\mu( d\zeta \mid E_\Lambda )(\omega) = \rho_\Lambda( \zeta \mid \omega ) \exp [ - E_\Lambda(\zeta \mid \omega) ] \pi(d\zeta) \quad \text{on} \quad F_\Lambda$$

(2.56 b) $$\rho_\Lambda( \cdot \mid \omega ) = 0 \quad \pi\text{ - a. s.} \quad \text{on} \quad \{ N(\Lambda) \neq \omega(\Lambda) \}$$

(2.56 c) $\rho_\Lambda( \cdot \mid \omega )$ is constant on each connected component of

$$\{ \alpha \subset \Lambda \;:\; \alpha(\Lambda) = \omega(\Lambda) , \; \alpha\, \omega_{S\setminus\Lambda} \in \Omega_r \} .$$

To this end, we fix a regular version of $\mu( \cdot \mid E_\Lambda )$ and an $\omega$ such that $\mu( \cdot \mid E_\Lambda )(\omega)$ is concentrated on $\{ \zeta \in \Omega_r \;:\; \zeta(\Lambda) = \omega(\Lambda), \; \zeta_{S\setminus\Lambda} = \omega_{S\setminus\Lambda} \}$ and so that (2.55) holds for all $C^1$ - functions $\varphi$ and $\psi$. We put $N = \omega(\Lambda)$ and

$$m = \exp H_{\Lambda,\omega} \quad \mu( \cdot \mid E_\Lambda )(\omega) \mid F_\Lambda .$$

Alternatively, we think of $m$ as a symmetric measure on the open set

$$U = \{ (x_1,\ldots,x_N) \in \Lambda^N \;:\; x_1 \ldots x_N \, \omega_{S\setminus\Lambda} \in \Omega_r \}$$

defined by

$$\int F( x_1,\ldots,x_N ) \; m( dx_1,\ldots,dx_N ) = \int F_{sym}(\zeta_\Lambda) \exp H_{\Lambda,\omega}(\zeta) \, \mu( d\zeta \mid E_\Lambda )(\omega).$$

Here $F$ is any nonnegative measurable function on $U$ and

$$F_{sym}(\alpha) = \begin{cases} \frac{1}{N!} \sum_{\tau} F( x_{\tau(1)},\dots,x_{\tau(N)} ) & \text{if} \quad \alpha = \{ x_1,\dots,x_N \} \subset \Lambda \\ 0 & \text{otherwise} \end{cases} ,$$

the sum extending over all permutations $\tau$ of $\{ 1,\dots,N \}$ . $m$ is a Radon measure on $U$ since $\Phi$ is continuous and hence $H_{\Lambda,\omega}$ is bounded on compact subsets of $U$ .

We claim that (2.56) will be proved once we have shown that on each connected component of $U$ $m$ is a multiple of $\lambda^N$ . Indeed, the preimage of each connected component of $\{ \alpha \subset \Lambda \ : \ \alpha(\Lambda) = N, \ \alpha\, \omega_{S\setminus\Lambda} \in \Omega_r \}$ under the mapping $( x_1,\dots,x_N) \to \{ x_1,\dots,x_N \}$ is the union of connected components of $U$ which can be obtained from a single such component by a permutation of the coordinates. Thus the claim follows from the symmetry of $m$ .

We may therefore confine our considerations to a fixed connected open subset $M$ of $U$ . Moreover, we may assume that the images of $M$ under all permutations of the coordinates are mutually disjoint. This is because otherwise we can represent $M$ as a union of such regions, namely

$$M = \bigcup_{a,\tau} M \cap U_{a,\tau} ,$$

where the union is taken over all $a \in \mathbb{R}^d$ and all permutations $\tau$ of $\{ 1,\dots,N \}$, and

$$U_{a,\tau} = \{ ( q_1, p_1,\dots,q_N, p_N ) \in U \ : \ a\, q_{\tau(1)} < \dots < a\, q_{\tau(N)} \} .$$

Thus if $m$ is absolutely continuous with respect to $\lambda^N$ on each of the sets $M \cap U_{a,\tau}$ then this is also true on the whole of $M$ , and if the density is constant on each $M \cap U_{a,\tau}$ then by symmetry it is constant on each of the overlapping sets $\bigcup_{\tau} M \cap U_{a,\tau}$ and hence is constant everywhere.

Now we let $F$ denote an arbitrary $C^\infty$ - function on $U$ with compact support in $M$ . We fix two indices $1 \leq i \leq 2d$ , $1 \leq j \leq N$ , and choose a $C^\infty$ - function

$G$ with compact support in $M$ such that $G(x_1,\dots,x_N) = x_j^i$ on the support of $F$ ; here $x_j^i$ denotes the i-th coordinate of $x_j = (x_j^1,\dots,x_j^{2d})$ . Let

$$\varphi(\alpha) = \begin{cases} F_{sym}(\alpha) & \text{if } \alpha \subset \Lambda\,,\ \alpha(\Lambda) = N\,,\ \alpha\,\omega_{S\setminus\Lambda} \in \Omega_r \\ 0 & \text{otherwise}\,, \end{cases}$$

and

$$\psi(\alpha) = \begin{cases} G_{sym}(\alpha)\,\exp H_{\Lambda,\omega}(\alpha) & \text{if } \alpha \subset \Lambda\,,\ \alpha(\Lambda) = N\,,\ \alpha\,\omega_{S\setminus\Lambda} \in \Omega_r \\ 0 & \text{otherwise}\,. \end{cases}$$

Then $\varphi$ and $\psi$ satisfy (2.51 a,b) except that $\psi$ is possibly only of class $C^1$ . Furthermore, for all $\zeta \in \Omega_r$ with $\zeta(\Lambda) = N$ and $\zeta_\Lambda\,\omega_{S\setminus\Lambda} \in \Omega_r$ we have $\tilde{\varphi}(\zeta) = F_{sym}(\zeta_\Lambda)$ and $\tilde{\psi}(\zeta)\exp[-H_{\Lambda,\omega}(\zeta)] = G_{sym}(\zeta_\Lambda)$ . Moreover, since the images of $M$ under permutations of the coordinates are assumed to be mutually disjoint,

$$\{F_{sym}, G_{sym}\} = \{F, G\}_{sym}\,,$$

where $\{F, G\}_{sym}$ denotes the symmetrization of

$$\{F, G\} = \sum_{j=1}^{N}\left[\frac{\partial F}{\partial q_j}\frac{\partial G}{\partial p_j} - \frac{\partial F}{\partial p_j}\frac{\partial G}{\partial q_j}\right]\,.$$

Therefore (2.55) gives us

$$0 = \int \{\tilde{\varphi}, \tilde{\psi}\,\exp[-H_{\Lambda,\omega}]\}\,\exp H_{\Lambda,\omega}\;d\mu(\,\cdot \mid E_\Lambda)(\omega)$$

$$= \int \{F, G\}\;dm\,.$$

Now observe that by the definition of $G$

$$\{ F, G \} = \begin{cases} -\dfrac{\partial F}{\partial x_j^{i+d}} & \text{if} \quad i \leq d \\[2ex] \dfrac{\partial F}{\partial x_j^{i-d}} & \text{if} \quad i > d \end{cases} .$$

Thus we have shown that if $m$ is considered as a distribution on $M$ then all its partial derivatives vanish. Since $M$ is open and connected, a well-known theorem (see, for instance, section 1.7.3 of Shilov (1968) ) implies $m$ on $M$ is a multiple of $\lambda^N$ . This completes the proof of (2.56) .

3. The *final step* in the proof of the theorem is to deduce from (2.56) that $\mu \in C$ . This is evident if $r = 0$ since in this case the set $\{ \alpha \subset \Lambda : \alpha(\Lambda) = \omega(\Lambda) , \alpha\, \omega_{S\setminus\Lambda} \in \Omega_r \}$ is always connected. In the case $r > 0$ , we have to use the condition that $\mu$ is loosely packed. Fix a connected set $\Lambda \in S_q$ , and for any connected $\Delta \supset \Lambda$ let $G(\Delta)$ denote the set of all those $\omega \in \Omega_r$ such that

$$\{ \zeta\, \omega_{\Delta\setminus\Lambda} : \zeta \subset \Lambda , \zeta(\Lambda) = \omega(\Lambda) , \zeta\, \omega_{S\setminus\Lambda} \in \Omega_r \}$$

is contained in a connected subset of

$$\{ \alpha \subset \Delta : \alpha(\Delta) = \omega(\Delta) , \alpha\, \omega_{S\setminus\Delta} \in \Omega_r \} .$$

Then $G(\Delta) \in E_\Lambda$ , and by assumption $G(\Delta) \uparrow \Omega_r$ $\mu$ - a. s. when $\Delta \uparrow S$ . Now we choose an $N \geq 0$ , an $F_\Lambda$- measurable set $A \subset \{ N(\Lambda) = N \}$ , an $E_\Lambda$ - measurable set $B \subset \{ N(\Lambda) = N \}$ , and put $B(\Delta) = B \cap G(\Delta)$ . Then we can write, using (2.56) and (1.56) ,

$$\int_{B(\Delta)} 1_A \, d\mu$$
$$= \int \mu(d\omega) \int_A \pi(d\zeta) \exp [ - E_\Delta( \zeta | \omega ) ] \rho_\Delta( \zeta | \omega ) \; 1_{B(\Delta)} ( \zeta_\Delta\, \omega_{S\setminus\Delta} )$$

$$= \int \mu(d\omega) \int \pi(d\eta) \; Z_{\Lambda,N}( \eta_\Delta \, \omega_{S\setminus\Delta} ) \exp [ - E_\Delta( \eta_{\Delta\setminus\Lambda} | \omega ) ]$$

$$\int_A \pi(d\zeta) \, \gamma_{\Lambda,N}( \zeta | \zeta_\Lambda \, \eta_{\Delta\setminus\Lambda} \, \omega_{S\setminus\Delta} ) \, \rho_\Delta( \zeta_\Lambda \, \eta_{S\setminus\Lambda} | \omega ) \; 1_{B(\Delta)}( \zeta_\Lambda \, \eta_{\Delta\setminus\Lambda} \, \omega_{S\setminus\Delta} )$$

$$= \int \mu(d\omega) \int_{\{N(\Lambda)=N\}} \pi(d\xi) \exp [ - E_\Delta( \xi | \omega ) ]$$

$$\int_A \pi(d\zeta) \, \gamma_{\Lambda,N}( \zeta | \zeta_\Lambda \, \xi_{\Delta\setminus\Lambda} \, \omega_{S\setminus\Delta} ) \, \rho_\Delta( \zeta_\Lambda \, \xi_{S\setminus\Lambda} | \omega ) \; 1_{B(\Delta)}( \zeta_\Lambda \, \xi_{\Delta\setminus\Lambda} \, \omega_{S\setminus\Delta} ) \quad .$$

Observe now that $\zeta_\Lambda \, \xi_{\Delta\setminus\Lambda} \, \omega_{S\setminus\Delta} \in B(\Lambda)$ iff $\xi_\Delta \, \omega_{S\setminus\Delta} \in B(\Delta)$ , and that in this case

$$\rho_\Delta ( \zeta_\Lambda \, \xi_{S\setminus\Lambda} | \omega ) = \rho_\Delta ( \xi | \omega ) \quad .$$

Therefore the last expression equals

$$\int_{B(\Delta)} \mu(d\omega) \int_A \pi(d\zeta) \, \gamma_{\Lambda,\omega(\Lambda)} ( \zeta | \omega ) \quad .$$

Letting $\Delta \uparrow S$ we see that a version of $\mu( \cdot \mid E_\Lambda )$ is given by the canonical Gibbs distributions. By a similar argument the same is true when $\Lambda \in S_q$ is not necessarily connected. The proof is thus complete. ⌟

We conclude this section with the

(2.57) Proof of Proposition (2.29) :

" (a) ⇒ (b) " Clearly we need only to consider functions of the form

$$F( \zeta, \eta, \omega ) = \varphi(\zeta) \; \psi(\eta) \; g(\omega) \; h(\omega)$$

where $\varphi \geq 0$ and $\psi \geq 0$ vanish outside of a set of the form $\{N(\Lambda) = N(S) = N\}$, and $g \geq 0$ and $h \geq 0$ are measurable with respect to $F_\Lambda$ and $F_{S \setminus \Lambda}$, respectively. Then we have to show that

$$\int \mu(d\omega)\ h(\omega) \int \omega^{(N)}(d\zeta)\ \varphi(\zeta)\ g(\omega \setminus \zeta) \int \sigma^{(N)}(d\eta)\ e^{-E(\eta|\omega\setminus\zeta)}\ \psi(\eta)$$

is symmetric in $\varphi$ and $\psi$. This, however, is immediately seen by taking conditional expectations with respect to $E_\Lambda$ and using the identity

$$\int \pi(d\omega) \int \omega^{(N)}(d\zeta)\ G(\zeta, \omega \setminus \zeta)$$

$$= \int \pi(d\omega) \int \sigma^{(N)}(d\zeta)\ G(\zeta, \omega)$$

which follows from (1.57) by induction on $N$.

" (b) ⇒ (a) " For given $\Lambda \in S$ and $N \geq 0$, we let $f$ and $h$ denote two nonnegative functions on $\Omega$ vanishing outside of $\{N(\Lambda) = N\}$ and such that $f$ is $F_\Lambda$ - measurable and $h$ is $E_\Lambda$ - measurable. We then define a measurable function $F$ on $\Omega \times \Omega \times \Omega$ by putting

$$F(\zeta, \eta, \omega) = f(\zeta)\ 1_{\{N(\Lambda)=N\}}(\zeta)\ h(\zeta\,\omega)\ e^{-\sigma(\Lambda)} / Z_{\Lambda,N}(\omega)$$

if $\zeta \cap \omega = \emptyset$ and $= 0$ otherwise. With this $F$ (2.30) reduces to

$$\text{(2.58)} \qquad \int \mu(d\omega)\ h(\omega)\ f(\omega) = \int \mu(d\omega)\ h(\omega) \int \pi(d\eta)\ \gamma_{\Lambda,\omega(\Lambda)}(\eta|\omega)\ f(\eta) \quad .$$

This comes from the simple facts that (i) if $\omega(\Lambda) = N$, $\zeta \subset \omega$, $\zeta(\Lambda) = \zeta(S) = N$ then $\zeta = \omega_\Lambda$, (ii) if $(\eta\omega \setminus \zeta)(\Lambda) = N$, $\zeta \subset \omega$, $\zeta(\Lambda) = \zeta(S) = N$, $\eta(\Lambda) = \eta(S) = N$ then $\omega(\Lambda) = N$, $\zeta = \omega_\Lambda$, and $h(\eta\omega \setminus \zeta) = h(\omega)$, and (iii) if $\eta \subset \Lambda$ then $E(\eta|\omega \setminus \omega_\Lambda) = E_\Lambda(\eta|\omega)$. Since $\Lambda$, $N$, $f$, and $h$ are arbitrarily chosen, (2.58) implies that $\mu \in C$.

" (b) ⇒ (c) " There is nothing to prove.

" (c) ⇒ (b) " Suppose that $r(\Phi) = 0$ . Then (b) is equivalent to the statement that for all measurable functions $F \geq 0$ on $\Omega^3$ and all $N \geq 1$

$$\int \mu(d\omega) \int \omega^{(N)}(d\zeta) \int \sigma^{(N)}(d\eta)\; F(\zeta, \eta, \omega \setminus \zeta) \tag{2.59}$$

$$= \int \mu(d\omega) \int \omega^{(N)}(d\zeta) \int \sigma^{(N)}(d\eta)\; e^{E(\zeta|\omega\setminus\zeta) - E(\eta|\omega\setminus\zeta)}\; F(\eta, \zeta, \omega \setminus \zeta),$$

and (c) is equivalent to the same assertion for $N = 1$ . Thus we need only to prove an induction step from $N - 1$ to $N$ . This is based on the obvious recursion formulas

$$N \int \omega^{(N)}(d\zeta)\; f(\zeta) = \int \omega^{(N-1)}(d\zeta) \int (\omega \setminus \zeta)(dx)\; f(x\zeta)$$

$$= \int \omega(dx) \int (\omega \setminus x)^{(N-1)}(d\zeta)\; f(x\zeta)$$

$$N \int \sigma^{(N)}(d\zeta)\; f(\zeta) = \int \sigma^{(N-1)}(d\zeta) \int \sigma(dx)\; f(x\zeta) \quad ,$$

which are valid for arbitrary measurable functions $f \geq 0$ . By means of these relations the l. h. s. of (2.59) can be written as

$$N^{-2} \int \mu(d\omega) \int \omega^{(N-1)}(d\zeta) \int \sigma^{(N-1)}(d\eta)\; G(\zeta, \eta, \omega \setminus \zeta) \quad ,$$

where

$$G(\zeta, \eta, \omega) = \int \omega(dx) \int \sigma(dy)\; F(x\zeta, y\eta, \omega \setminus x) \quad .$$

By the induction hypothesis this equals

$$N^{-2} \int \mu(d\omega) \int \omega^{(N-1)}(d\zeta) \int \sigma^{(N-1)}(d\eta)\; e^{E(\zeta|\omega\setminus\zeta) - E(\eta|\omega\setminus\zeta)}$$

$$\int (\omega \setminus \zeta)(dx) \int \sigma(dy)\; F(x\eta, y\zeta, \omega \setminus x\zeta)$$

$$= N^{-2} \int \mu(d\omega) \int \omega(dx) \int \sigma(dy) \; G_1(\, x, y, \omega \setminus x \,) \quad ,$$

where $G_1(\, x, y, \omega \,) \quad =$

$$= \int \omega^{(N-1)} (d\zeta) \int \sigma^{(N-1)} (d\eta) \; e^{E(\zeta | x\omega\setminus\zeta) - E(\eta | x\omega\setminus\zeta)} \; F(\, x\eta, y\zeta, \omega \setminus \zeta \,) \quad .$$

Thus we get from (2.31) that the l. h. s. of (2.59) coincides with

$$N^{-2} \int \mu(d\omega) \int \omega(dx) \int \sigma(dy) \; e^{E(x|\omega\setminus x) - E(y|\omega\setminus x)}$$

$$\int (\omega \setminus x)^{(N-1)} (d\zeta) \int \sigma^{(N-1)} (d\eta) \; e^{E(\zeta | y\omega\setminus x\zeta) - E(\eta | y\omega\setminus x\zeta)} \; F(y\eta, x\zeta, \omega\setminus x\zeta)$$

This expression is exactly the r. h. s. of (2.59) . Indeed, it follows from the definition (2.28) that

$$E(\, x | \omega \setminus x \,) - E(\, y | \omega \setminus x \,) = E(\, x\,\zeta | \omega \setminus x\,\zeta \,) - E(\, y\,\zeta | \omega \setminus x\,\zeta \,)$$

and

$$E(\, \zeta | y\omega \setminus x\zeta \,) - E(\, \eta | y\omega \setminus x\zeta \,) = E(\, y\,\zeta | \omega \setminus x\,\zeta \,) - E(\, y\,\eta | \omega \setminus x\,\zeta \,) \, .$$

Thus the induction step is complete. ┘

*Bibliographical notes*: Particle jump processes of the type considered in section 2.1 were introduced by F. Spitzer (1970) . Theorem (2.14) generalizes a result of K. G. Logan (1974) . The proof of Proposition (2.38) gives a simplified version of ideas of R. Lang (1977) which have their origin in a paper of A. N. Kolmogorov (1937) . Theorem (2.53) is a slight modification of a result of Aizenman et al. (1977) who use an algebra being somewhat different from $\underline{\underline{A}}$ . Finally we want to mention that time evolutions of an Ornstein-Uhlenbeck type (which are, in a certain sense, intermediate between the two evolutions considered in section 2.2 ) are studied by J. Fritz (1978) .

## § 3 Spatially homogeneous models

In this section we discuss the discrete model (introduced in section 1.1) for the particular situation when the set $S$ of sites is the $d$ - dimensional integer lattice and the interaction potential is translation invariant. Moreover, we restrict our attention to those canonical Gibbs measures which are shift-invariant. These will be characterized by a variational principle in subsection 3.1. As a consequence we get the result that each shift-invariant canonical Gibbs measure $\mu$ is a mixture of grand canonical Gibbs measures; furthermore, the distribution of the activities in the representation of $\mu$ depends only on the distribution of the particle density corresponding to $\mu$. In subsection 3.2 we will show that the variational principle has a dynamical interpretation in terms of the time evolutions considered in section 2.1.

### 3.1 The variational principle

Let $S = \mathbb{Z}^d$ be the cubic lattice of dimension $d \geq 1$, $F$ a finite set of types of particles, and $\Omega = F^S$. We fix a shift-invariant potential $\Phi$, i.e., a potential $\Phi$ satisfying (1.35). Moreover, we assume that

$$(3.1) \qquad \|\Phi\| = \sum_{A \ni 0} \|\Phi(A, \cdot)\| < \infty$$

(cf. (1.13) ). Here $0$ denotes the origin in the lattice. Furthermore, without loss of generality we will assume that the chemical potential $\Phi(0, \cdot)$ is zero. Indeed, if this were not the case then we could replace $z(a)$ by $z(a)\, e^{-\Phi(0,a)}$ in the definition (1.16) of the grand canonical Gibbs distribution, and from definition (1.24) it is seen that canonical Gibbs distributions do not depend on the chemical potential whenever this is shift-invariant.

In this section, the notation $\Lambda \uparrow S$ will always have the meaning that $\Lambda$ runs through a sequence of cubes, i.e., of sets of the form

$$\{ x \in S \ : \ a \leq x^{(i)} \leq b \ \ (1 \leq i \leq d) \}$$

where $a, b \in \mathbb{Z}$ and $x^{(i)}$ is the i-th coordinate of $x$ .

We are going to prove a variational characterization of the set $C_\theta = C_\theta(\Phi)$ of all shift-invariant canonical Gibbs measures with respect to $\Phi$ . But first let us recall the variational principle for the set $G_\theta(z)$ of Gibbs measures which is due to Lanford and Ruelle (1969) .

Suppose that $\mu$ is a shift-invariant probability measure on $(\Omega, F)$ . Then the *specific entropy*

$$(3.2) \qquad h(\mu) = - \lim_{\Lambda \uparrow S} |\Lambda|^{-1} \int \mu(d\omega) \log \mu( X_V = \omega_V )$$

of $\mu$ exists as well as the *specific energy*

$$(3.3) \qquad e(\mu) = e(\mu, \Phi) = \int \mu(d\omega) \sum_{A \ni 0} \Phi(A, \omega)/|A|$$

$$= \lim_{\Lambda \uparrow S} |\Lambda|^{-1} \int \mu(d\omega) E_\Lambda(\omega_\Lambda | \alpha)$$

of $\mu$ . Moreover, for each $z \in A$ the *pressure* (which is sometimes also called the *specific free Gibbs-energy*)

$$(3.4) \qquad P(z) = P(z, \Phi) = \lim_{\Lambda \uparrow S} |\Lambda|^{-1} \log Z_\Lambda (z, \alpha)$$

exists. Here, as well as in (3.3) , $\alpha$ may run through an arbitrary sequence in $\Omega$ , and the limit does not depend on the choice of this sequence. The function $\mu \to e(\mu)$ is continuous since $\sum_{A \ni 0} \Phi(A, \cdot)/|A|$ is continuous, and the function $\mu \to h(\mu)$ is upper semicontinuous and affine. Furthermore, the expression

$$(3.5) \qquad g(\mu, z) = P(z) + e(\mu) - h(\mu) - \int \mu(d\omega) \log z(\omega_0)$$

is always nonnegative. $g(\mu, z) - P(z)$ is called the *specific free energy* of $\mu$ with respect to $\Phi$ and $z$. A proof of all the facts stated above can be found, for instance, in Föllmer (1973) and Ruelle (1969). For the proof of (3.4) see also stage 5 in (3.33) below.

The following variational principle of Lanford and Ruelle states that the affine and lower semicontinuous function $\mu \to g(\mu, z)$ attains its minimum exactly on the set $G_\theta(z)$.

(3.6) Theorem: *Suppose that* $z \in A$ *and* $\mu$ *is a shift-invariant probability measure on* $(\Omega, F)$. *Then* $\mu \in G(z)$ *iff* $g(\mu, z) = 0$.

For the proof we refer to Lanford/Ruelle (1969) and Preston (1976). (Note that we may assume without loss of generality that $\prod_{a \in F} z(a) > 0$. Indeed, if $F_z = \{a \in F : z(a) > 0\}$ then each of the statements "$\mu \in G(z)$" and "$g(\mu, z) = 0$" implies that $\mu((F_z)^S) = 1$.)

Now we introduce certain "canonical" quantities $Q(\rho)$ and $c(\mu)$ which are analoguous to the "grand canonical" quantities $P(z)$ and $g(\mu, z)$. We let

$$(3.7) \qquad A_1 = \{ \rho \in A : \sum_{a \in F} \rho(a) = 1 \}$$

denote the set of all probability vectors on $F$. For the sake of simplicity we confine our attention to a particular sequence of cubes, namely

$$(3.8) \qquad \Lambda_n = \{ x \in S : -2^n < x^{(i)} \leq 2^n \quad (1 \leq i \leq d) \}$$

$(n \geq 1)$ (Due to the homogeneity of the model it does not matter whether $\Lambda_n$ is replaced by any other cube of side $2^{n+1}$.)

(3.9) Proposition: *Suppose that* $\rho \in A_1$, $L_n \in A_{\Lambda_n}$ *with* $L_n/|\Lambda_n| \to \rho (n \to \infty)$ *and* $\alpha_n \in \Omega$ *are given. Then the limit*

$$(3.10) \qquad Q(\rho) = Q(\rho, \Phi) = \lim_{n \to \infty} |\Lambda_n|^{-1} \log Z_{\Lambda_n, L_n}(\alpha_n)$$

*exists and depends only on* $\rho$ *(and* $\Phi$ *). The function* $Q(\,\cdot\,)$ *on* $A_1$ *is concave, and for all* $z \in A$ *we have*

$$(3.11) \qquad P(z) = \max_{\rho \in A_1} [\, Q(\rho) + \rho \log z \,] \; .$$

*Moreover, for each* $\rho \in A_1$ *there is a unique* $z = z(\rho) \in A_1$ *solving the equation*

$$(3.12) \qquad P(z) - \rho \log z - Q(\rho) = 0 \; .$$

In order that the scalar product

$$(3.13) \qquad \rho \log z = \sum_{a \in F} \rho(a) \log z(a)$$

is always well-defined we use the convention $0 \log 0 = 0$ . $Q(\rho)$ is called the *specific free Helmholtz energy per volume.*

Proposition (3.9) , as well as the analoguous result for continuous systems, has a long history associated with the names of L. van Hove, D. Ruelle, M.E. Fisher, R.L. Dobrushin and R.A. Minlos; see, for instance, Dobrushin/Minlos (1967) , lecture iv in Minlos (1968) , Theorem 3.4.6 in Ruelle (1969) , and Lanford (1973) . For lattice systems a more general but weaker version of (3.9) is contained in Thompson (1974) . For the convenience of the reader we will give a proof of (3.9) at the end of section 3.1 .

Now we consider the function $\rho \to z(\rho) = (\, z(a, \rho) \,)_{a \in F}$ in more detail. It follows immediately from (3.12) and (3.11) that the faces of the simplex $A_1$ and its interior

$$A_1^0 = \{\, \rho \in A_1 \; : \; \prod_{a \in F} \rho(a) > 0 \,\}$$

remain invariant under the mapping $\rho \to z(\rho)$ . Furthermore, observe that $z$ is a solution of (3.12) with a given $\rho \in A_1^0$ iff the plane $\rho' \to -\rho' \log z$ is a tangent plane of $Q(\cdot)$ at $\rho$. Thus we can reformulate the last sentence of (3.9) by saying that the concave function $Q(\cdot)$ has a unique tangent plane at each point $\rho \in A_1^0$. (On the faces of $A_1$ a similar statement is true, of course.) Fixing some $b \in F$ and using $\rho(a)$ $(a \neq b)$ as local coordinates on $A_1^0$ we get the result: For all $\rho \in A_1^0$ and $a \neq b$ the partial derivative $\frac{\partial}{\partial \rho(a)} Q(\rho)$ exists, is a continuous function of $\rho$ (by convexity), and satisfies

$$(3.14) \qquad \frac{\partial}{\partial \rho(a)} Q(\rho) + \log \frac{z(a,\rho)}{z(b,\rho)} = 0 .$$

Thus for any $a \in F$ we have

$$(3.15) \qquad z(a, \rho) = \exp\left[ - \frac{\partial}{\partial \rho(a)} Q(\rho) \right] / \sum_{c \in F} \exp\left[ - \frac{\partial}{\partial \rho(c)} Q(\rho) \right] .$$

In particular, the function $\rho \to z(\rho)$ is continuous on $A_1^0$. If $\Phi$ satisfies the condition

$$\sum_{A \ni 0} ( |A| - 1 ) \, \| \Phi(A, \cdot) \| < \infty$$

then it can be shown (using the methods of Dobrushin/Minlos (1967) ; we give an explicit proof in (7.4) ) that

$$(3.16) \qquad - \frac{\partial}{\partial \rho(a)} Q(\rho) = \lim_{n \to \infty} \log Z_{\Lambda_n, L_n + 1_a - 1_b}(\alpha_n) / Z_{\Lambda_n, L_n}(\alpha_n)$$

whenever $L_n / |\Lambda_n| \to \rho$ and $\alpha_n \in \Omega$. Here $1_a$ denotes the indicator function of the set $\{a\} \subset F$. Together with (5.28) it follows that for all $a, b \in F$

$$(3.17) \qquad \left| \log \frac{\rho(a)}{\rho(b)} - \log \frac{z(a,\rho)}{z(b,\rho)} \right| \leq 2 \| \Phi \| .$$

Furthermore, (3.14) and (3.16) imply

$$(3.18) \qquad \frac{z(b,\rho)}{z(a,\rho)} = \lim_{n\to\infty} Z_{\Lambda_n, L_n + 1_a - 1_b}(\alpha_n) \,/\, Z_{\Lambda_n, L_n}(\alpha_n) \quad .$$

A suitable extension of this result will be the key to our considerations in § 5 (see (5.14) , in particular (b) and (d) ) .. Finally, let us remark that the mapping $\rho \to z(\rho)$ is in general not invertible, i.e., that $Q(\cdot)$ is in general not strictly concave. It is known that $|G_\theta(z)| > 1$ as soon as $|\{z(\cdot) = z\}| > 1$ and that for attractive potentials the converse is also true. Examples of this phenomenon (which has the physical interpretation of a phase transition) are given in section 5.3 of Ruelle (1969) , for instance.

In the case $\Phi \equiv 0$ a simple application of Stirling's formula shows that $Q(\rho) = -\rho \log \rho$ and $P(z) = \log \sum_{a\in F} z(a)$ . Hence $z(\rho) = \rho$ .

In order to find the "canonical counterpart" of the function $g(\mu, z)$ let us first give a variational characterization of Gibbs measures with a fixed particle density:

(3.19) Remark: *Suppose that* $\rho \in A_1$ *and* $\mu$ *is a shift-invariant probability measure on* $(\Omega, F)$ *such that* $\mu(X_0 = a) = \rho(a)$ $(a \in F)$ . *Then*

$$h(\mu) - e(\mu, \Phi) \le Q(\rho, \Phi)$$

*with equality iff* $\mu \in G_\theta(\, z(\rho), \Phi\, )$ .

Proof: Combine (3.5) , (3.12) , and (3.6) . ┘

If $\mu$ is shift-invariant and $a \in F$ then the $d$ - dimensinal version of Birkhoff's ergodic theorem (see, for instance, Theorem VIII. 6. 9 in Dunford/Schwartz (1958) ) shows that for $\mu$ - a. e. $\omega$ the density

$$(3.20) \qquad \rho(a, \omega) = \lim_{\Lambda\uparrow S} N(a, \omega_\Lambda) \,/\, |\Lambda|$$

of a - particles exists. Clearly, $\rho(\omega) = \rho(\cdot, \omega) \in A_1$, and if $\mu$ is ergodic then $\rho(\cdot)$ is a. s. constant. Thus (3.19) shows that for ergodic $\mu$

$$e(\mu) - h(\mu) + \int Q(\rho(\cdot))\, d\mu \geq 0 \quad .$$

Representing any shift-invariant $\mu$ as a mixture of ergodic probability measures we obtain the same inequality for all shift-invariant probability measures $\mu$. (A direct proof will be given in (3.23) ). Furthermore, (3.19) shows that equality holds for all $\mu \in \bigcup_z \text{ex}\, G_\theta(z)$. Thus the function

$$\text{(3.21)} \qquad \begin{aligned} c(\mu) &= c(\mu, \Phi) \\ &= e(\mu, \Phi) - h(\mu) + \int \mu(d\omega)\, Q(\rho(\omega), \Phi) \end{aligned}$$

is a good candidate for a function which attains its minimum exactly on $C_\theta$. The next remark implies that $c(\cdot)$ attains its minimum on a compact face of the simplex of all shift-invariant probability measures.

(3.22) Remark: *The function* $\mu \to c(\mu)$ *is affine and lower semicontinuous.*

Proof: We need only show that the function $\mu \to \int Q(\rho(\cdot))\, d\mu$ is lower semi-continuous. Let $\rho_n(\omega) = N(\omega_{\Lambda_n})/|\Lambda_n|$, where $\Lambda_n$ is defined by (3.8). Filling $\Lambda_{n+1}$ by $2^d$ translates of $\Lambda_n$ and using the concavity of $Q$ as well as the shift-invariance of $\mu$ we obtain the inequality

$$\int Q(\rho_{n+1}(\cdot))\, d\mu \geq \int Q(\rho_n(\cdot))\, d\mu$$

for all $n \geq 1$. Hence $\mu \to \int Q(\rho(\cdot))\, d\mu$ is the limit of an increasing sequence of continuous functions. ⌟

The next proposition shows that $c(\mu)$ is a specific conditional information gain.

(3.23) Proposition: *Suppose that* $\mu$ *is shift-invariant and* $(\alpha_n)_{n\geq 1}$ *is an arbitrary sequence in* $\Omega$ . *Then*

(3.24) $\quad c(\mu) =$

$$= \lim_{n\to\infty} |\Lambda_n|^{-1} \int \mu(d\omega) \log [\, \mu(X_{\Lambda_n} = \omega_{\Lambda_n} | N(X_{\Lambda_n}) = N(\omega_{\Lambda_n}))/\gamma_{\Lambda_n, N(\omega_{\Lambda_n})}(\omega_{\Lambda_n}|\alpha_n) \,] \,.$$

*In particular,* $c(\mu) \geq 0$ . *For* $\mu \in C_\theta$ *we have* $c(\mu) = 0$ .

The proof is postponed until (3.29) . There is a simple relation between $c(\mu)$ and $g(\mu, z)$ which is closely connected with the Legendre-transformation (3.11) : For any $z \in A$

(3.25) $\quad g(\mu, z) - c(\mu) =$

$$= \int \mu(d\omega) \;[\, P(z) - \rho(\omega) \log z - Q(\rho(\omega))\,] \;\geq\; 0 \;.$$

Together with the "grand canonical variational principle" (3.6) this yields the following "canonical variational principle" :

(3.26) <u>Theorem:</u> *Suppose that* $\mu$ *is a shift-invariant probability measure on* $(\Omega, F)$ *Then* $\mu \in C_\theta$ *iff* $c(\mu) = 0$ .

<u>Proof:</u> It is sufficient to show that $c(\mu) = 0$ implies $\mu \in C_\theta$ . Using (3.22) and the ergodic decomposition of $\mu$ we may assume that $\mu$ is ergodic. Then there is some $\rho \in A_1$ such that $\rho(\,.\,) = \rho$ $\mu$ - a. s. . For $z = z(\rho)$ the r. h. s. of (3.25) vanishes. Hence $g(\mu, z(\rho)) = 0$ , and (3.6) implies that $\mu \in G(z(\rho)) \subset C$ . ⌟

In section 3.2 we will give a second proof of this canonical variational principle (under a condition on $\Phi$ which is somewhat stronger than (3.1) ) .

(3.27) <u>Corollary:</u> *Let* $\Phi$ *be a shift-invariant potential satisfying* (3.1) . *Then*

(i) $\quad \text{ex}\, C_\theta \;=\; \bigcup_{z\in A} \;\text{ex}\, G_\theta(z)$

(ii) *Suppose that* $\mu \in C_\theta$ *and* $z \in A$ . *Then* $\mu \in G_\theta(z)$ *iff* $z(\rho(\cdot)) \sim z$ $\mu$ *- a. s.*

(iii) *Assume* $\mu \in C_\theta$ . *Let* m *denote the distribution of* $z(\rho(\cdot))$ *under* $\mu$ , *and* $(z, A) \to \mu_z(A)$ *a regular version of the conditional distribution* $\mu(A | z(\rho(\cdot)) = z)$ $(z \in A_1 , A \in F)$ . *Then*

$$m( z \in A_1 : \mu_z \in G_\theta(z) ) = 1$$

*and*

$$\mu = \int \mu_z \ m(dz) \quad .$$

Proof: "(ii)": If $\mu \in G_\theta(z)$ then the l. h. s. of (3.25) vanishes, and (3.9) implies that $z(\rho(\cdot)) \sim z$ $\mu$ - a. s. Conversely, if $\mu \in C_\theta$ and $z(\rho(\cdot)) \sim z$ $\mu$ - a. s. then (3.25) and (3.26) show that $g(\mu, z) = 0$ .

"(i)" Every $\mu \in \text{ex } C_\theta$ is ergodic. Hence we can find some $z(\mu) \in A_1$ such that $z(\rho(\cdot)) = z(\mu)$ $\mu$ - a. s. Therefore we get from (ii) that $\mu \in \bigcup_z \text{ex } G_\theta(z)$ . Conversely, each $\mu \in \bigcup_z \text{ex } G_\theta(z)$ is ergodic and therefore extreme in $C_\theta$ .

"(iii)" This follows from (ii) by standard arguments, see stage 8 of (5.27) or stage 4 of (6.27) . ┘

In the particular case $\Phi \equiv 0$ we have $C = C_\theta$ , $z(\rho) = \rho$ and $G(\rho) = \{\rho^S\}$ $(\rho \in A_1)$ . Thus we get the following

(3.28) Corollary (de Finetti's Theorem): *Suppose that* $\mu$ *is a symmetric (see* (1.29)) *probability measure on* $(\Omega, F)$ *and* m *is the distribution of* $\rho(\cdot)$ *under* $\mu$ . *Then*

$$\mu = \int_A \rho^S \ m(d\rho) \quad .$$

For a Markovian example see Georgii (1975) . We conclude this section with the proof of (3.23) and (3.9) .

(3.29) Proof of Proposition (3.23):

1. In order to prove (3.24) we write the integrals on the r. h. s. in the form

$$\int \mu(d\omega) \log \mu( X_\Lambda = \omega_\Lambda ) - \int \mu(d\omega) \log \mu( N(X_\Lambda) = N(\omega_\Lambda) )$$

$$+ \int \mu(d\omega) \; E_\Lambda( \omega_\Lambda | \alpha ) + \int \mu(d\omega) \log Z_{\Lambda, N(\omega_\Lambda)} (\alpha)$$

and divide each term by $|\Lambda|$ . Then the first term tends to $- h(\mu)$ . The second term tends to zero, since

$$0 \leq \sum_{L \in A_\Lambda} \mu( N(X_\Lambda) = L ) \log \mu( N(X_\Lambda) = L ) \leq \log |A_\Lambda|$$

and $\lim_{\Lambda \uparrow S} |\Lambda|^{-1} \log |A_\Lambda| = 0$ . By (3.3) the third term tends to $e(\mu)$. Finally, (3.10) implies that the fourth term tends to $\int \mu(d\omega)\, Q( \rho(\omega) )$ ; indeed, we may apply the dominated convergence theorem since

(3.30) $$e^{-|\Lambda| \, \| \Phi \|} \leq Z_{\Lambda,L} (\alpha) \leq |F|^{|\Lambda|} \, e^{|\Lambda| \, \| \Phi \|} .$$

2. The integrals on the r. h. s. of (3.24) can be written in the form

(3.31) $$\sum_{L \in A_\Lambda} \mu(N(X_\Lambda) = L) \sum_{\zeta \in \Omega_{\Lambda,L}} \mu(X_\Lambda = \zeta | N(X_\Lambda) = L) \log \frac{\mu(X_\Lambda = \zeta | N(X_\Lambda) = L)}{\gamma_{\Lambda,L} (\zeta|\alpha)} .$$

The inner sum has the form of an information gain which is known to be non-negative.

3. For $\mu \in C_\theta$ we have $c(\mu) = 0$ . Indeed for any $\zeta \in \Omega_\Lambda$ and $\omega, \alpha \in \Omega$

$$| E_\Lambda (\zeta|\omega) - E_\Lambda (\zeta|\alpha) | \leq 2 \; r(\Lambda) ,$$

where

$$(3.32) \qquad r(\Lambda) = \sum_{\substack{A\cap\Lambda\neq\emptyset \\ A\setminus\Lambda\neq\emptyset}} \| \Phi(A,\cdot) \| .$$

Thus for each $L \in A_\Lambda$ with $\mu(N(X_\Lambda) = L) > 0$ and each $\zeta \in \Omega_{\Lambda,L}$ the ratio

$$\mu(X_\Lambda = \zeta | N(X_\Lambda) = L) / \gamma_{\Lambda,L}(\zeta|\alpha)$$

$$= \int \mu(d\omega | N(X_\Lambda) = L)\, \gamma_{\Lambda,L}(\zeta|\omega) / \gamma_{\Lambda,L}(\zeta|\alpha)$$

lies between $e^{-4r(\Lambda)}$ and $e^{4r(\Lambda)}$. Therefore the expression (3.31) is bounded by $4\, r(\Lambda)$. The assertion follows now from the fact that $\lim_{\Lambda\uparrow S} r(\Lambda)/|\Lambda| = 0$. In order to see that this limit is zero we choose an $\varepsilon > 0$ and an $R < \infty$ such that

$$\sum_{A\ni 0:\mathrm{diam}\,A>R} \| \Phi(A,\cdot) \| < \varepsilon .$$

Denoting by $\Lambda_R$ the set of all $x \in \Lambda$ whose distance from $S \setminus \Lambda$ is less than $R$ we obtain the inequality

$$r(\Lambda) \leq |\Lambda|\,\varepsilon + |\Lambda_R|\, \| \Phi \| ,$$

and the claim follows since for a sequence of cubes $\lim_{\Lambda\uparrow S} |\Lambda_R|/|\Lambda| = 0$. ┘

(3.33) Proof of Proposition (3.9):

1. For the proof of the existence and concavity of $Q$ we may assume that $\Phi$ has a finite range $R < \infty$ ( i. e. that $\Phi(A) = 0$ unless $\mathrm{diam}\,A \leq R$ ). This is because the potentials of finite range are dense in the Banach space of all shift-invariant potentials with finite norm (3.1), and because we always have

$$\left|\ |\Lambda|^{-1} \log Z_{\Lambda,L}(\alpha,\Phi) - |\Lambda|^{-1} \log Z_{\Lambda,L}(\alpha,\Psi)\ \right| \leq \|\Phi - \Psi\| \quad .$$

2. For each $n \geq 1$ we define a function $Q_n$ on $A_1$ by putting

$$Q_n(\rho) = |\Lambda_n|^{-1} \log Z_{\Lambda_n,L}(\alpha_n)$$

if $\rho = L/|\Lambda_n|$ for some $L \in A_{\Lambda_n}$ and extending $Q_n$ to all other $\rho$ by affine interpolation. We use the abbreviation $D = 2^d$ . We claim that for any $\rho$ of the form $\rho = D^{-1} \sum_1^D \rho_i$

$$Q_{n+1}(\rho) \geq D^{-1} \sum_1^D Q_n(\rho_i) - 2\, r(\Lambda_n) / |\Lambda_n| \quad ,$$

where $r(\Lambda_n)$ is defined by (3.32) . It is sufficient to prove this inequality in the particular case when $\rho_i = L_i/|\Lambda_n|$ with $L_i \in A_{\Lambda_n}$ . To this end we fill $V = \Lambda_{n+1}$ by $D$ pairwise disjoint translates $V_1,\ldots,V_D$ of $\Lambda_n$ . Then the claim is equivalent to the estimate

$$Z_{V,\sum_1^D L_i}(\alpha) \geq \left( \prod_1^D Z_{V_i,L_i}(\alpha_i) \right) \exp\left[ -2\, D\, r(\Lambda_n) \right] \quad .$$

But this is obvious since for any $\zeta \in \Omega_V$

$$\left| E_V(\zeta|\alpha) - \sum_1^D E_{V_i}(\zeta_{V_i}|\alpha_i) \right| \leq \sum_1^D 2\, r(V_i) \quad .$$

3. For a finite range potential $\Phi$ we have

$$\sum_1^\infty r(\Lambda_n) / |\Lambda_n| < \infty \quad ,$$

and from 2. it follows that for any $\rho \in A_1$ the sequence

$$Q_n(\rho) \;-\; \sum_{n}^{\infty} 2\, r\,(\Lambda_j)\;/\;|\Lambda_j|$$

is increasing. By (3.30) this sequence is bounded. This implies that $Q_n(\rho)$ converges to some $Q(\rho)$ .

4. From 2. we further deduce the inequality

$$Q(\tfrac{1}{2}\rho_1 + \tfrac{1}{2}\rho_2) \;\geq\; \tfrac{1}{2}Q(\rho_1) + \tfrac{1}{2}Q(\rho_2)$$

$(\rho_1, \rho_2 \in A_1)$ which proves the concavity and continuity of $Q(\,\cdot\,)$ . In particular, we obtain from Dini's theorem that the convergence $Q_n \to Q$ is uniform on $A_1$ . Thus the proof of (3.10) is complete.

5. The existence of the pressure $P$ and the Legendre transformation (3.11) follow from the estimate

$$\max_{L\in A_\Lambda} e^{L \log z}\; Z_{\Lambda,L}(\alpha) \;\leq\; Z_\Lambda(z, \alpha) \;=$$

$$= \sum_{L\in A_\Lambda} \Big(\prod_{a\in F} z(a)^{L(a)}\Big)\, Z_{\Lambda,L}(\alpha) \;\leq\; |A_\Lambda| \max_{L\in A_\Lambda} e^{L \log z}\; Z_{\Lambda,L}(\alpha) \quad .$$

6. For each $\rho \in A_1$ we have

$$Q(\rho) \;=\; \inf_{z\in A} \,[\, P(z) - \rho \log z \,] \quad .$$

For suppose there is some $\rho_0 \in A_1$ and $\varepsilon > 0$ such that

$$Q(\rho_0) + \varepsilon \;\leq\; P(z) - \rho_0 \log z \qquad (z \in A) \quad .$$

Separating the convex sets

$$\{ (\rho, \lambda) \in A_1 \times \mathbb{R} \; : \; - \| \Phi \| \le \lambda \le Q(\rho) \}$$

and $\{\rho_0\} \times [\, Q(\rho_0) + \varepsilon, \infty\, [$ in $\mathbb{R}^F \times \mathbb{R}$ by a hyperplane we get some $w \in \mathbb{R}^F$ and $\alpha \in \mathbb{R}$ such that

$$\rho_0 w + P(z) - \rho_0 \log z > \alpha \qquad (z \in A)$$

$$\rho w + Q(\rho) < \alpha \qquad (\rho \in A_1) \quad .$$

For $z(a) = e^{w(a)}$ we obtain a contradiction to (3.11) .

7. For each $\rho \in A_1$ there is some $z = z(\rho) \in A_1$ solving (3.12) . Indeed, for any $c > 0$ we deduce from (3.11) that

$$P(z) - \rho \log z = P(c\, z) - \rho \log (c\, z) \quad .$$

Thus the $\inf\limits_{z \in A}$ in 6. can be replaced by $\inf\limits_{z \in A_1}$ . This proves the assertion since the lower semicontinuous function $z \to P(z) - \rho \log z$ attains its minimum on the compact set $A_1$ .

8. $z(\rho)$ is uniquely determined. For suppose there is a $\rho \in A_1$ and $z_1, z_2 \in A_1$ solving (3.12) . Define $z(a) = \sqrt{z_1(a)\, z_2(a)}$ . Then from (3.11) we obtain

$$\begin{aligned} Q(\rho) &\le P(z) - \rho \log z \\ &\le \tfrac{1}{2} [\, P(z_1) + \rho \log z_1 \,] + \tfrac{1}{2} [\, P(z_2) + \rho \log z_2 \,] \\ &= Q(\rho) \quad . \end{aligned}$$

This implies $P(z) = \frac{1}{2} P(z_1) + \frac{1}{2} P(z_2)$ and therefore $z_1 = z_2$ . Indeed, if $\mu \in G_\theta(z)$ then

$$0 = g(\mu, z) = \frac{1}{2} g(\mu, z_1) + \frac{1}{2} g(\mu, z_2)$$

and thus $\mu \in G_\theta(z_1) \cap G_\theta(z_2)$ . Therefore $\gamma_0^{z_1}(a| \cdot ) = \gamma_0^{z_2}(a| \cdot )$ $\mu$ - a. s. for each $a \in F$ . From this it follows immediately that $z_1 = z_2$ . ┘

## 3.2 The free energy as a function of time

We consider a particle jump process of the type introduced in section 2.1 . We have shown that the canonical Gibbs measures with respect to an interaction $\Phi$ are invariant (even time reversible) under such an evolution provided the jump rates are related to $\Phi$ by the condition $(R_\Phi)$ . Now we will prove that the specific free energy of the distribution at time t decreases as t increases.

We assume that we are given a rate function $c( \cdot , \cdot )$ satisfying the following properties:

(i) (Homogeneity) There is a function from $S \times \Omega$ to $[0, \infty[$ (which is also denoted by c ) such that

(3.34) $$c( xy, \theta_x \omega ) = c( y - x, \omega )$$

whenever $xy \subset S$ and $\omega \in \Omega$ .

(ii) There is a symmetric function $c : S \to \mathbb{R}$ and a constant $K > 0$ such that

(3.35) $$c( y - x ) \leq c( xy, \cdot ) \leq K c( y - x )$$

for all $xy \subset S$ .

(iii) The matrix $( c(y - x) )_{x,y \in S}$ is irreducible.

(iv) $c(\cdot,\cdot)$ satisfies condition $(R_\Phi)$ (see (2.11) ) where $\Phi$ is a shift-invariant potential of finite norm (3.1) .

Moreover, we assume that the conditions of the existence theorem (2.4) are satisfied. Then in particular we have

(3.36) $$\sum_{x \in S} c(x) < \infty .$$

We let again $(P_t)_{t \geq 0}$ denote the Markovian semigroup of transition operators on $C(\Omega)$ which is defined by the rate function $c(\cdot,\cdot)$ . This semigroup also acts on probability measures via the formula

$$\int f \, d(\mu P_t) = \int (P_t f) \, d\mu \qquad ( f \in C(\Omega) ) .$$

For the sake of brevity we will usually write $\mu_t$ instead of $\mu P_t$ . If $\mu$ is shift-invariant then the homogeneity of $c(\cdot,\cdot)$ implies that $\mu_t$ is also shift-invariant for every $t > 0$ ; this is an immediate consequence of the construction of $P_t$ , see, e. g. , Liggett (1971) and Sullivan (1975) . Therefore for any shift-invariant $\mu$ the function

$$t \to c(\mu_t) = e(\mu_t, \Phi) - h(\mu_t) + \int Q(\rho(\ ), \Phi) \, d\mu_t$$

on $[0, \infty[$ is well-defined. We are interested in the properties of this function.

First we observe that the term involving the integral is a constant function of $t$ . We let $\rho(\mu_t)$ denote the distribution of the particle density $\omega \to \rho(\omega)$ under $\mu_t$ .

(3.37) <u>Remark:</u> *Suppose that* $\mu$ *is a shift-invariant probability measure on* $(\Omega, F)$ *Then for each* $t > 0$ *we have* $\rho(\mu_t) = \rho(\mu)$ .

<u>Proof:</u> It is sufficient to show that

$$\int f(\rho(\cdot)) \, d\mu_t = \int f(\rho(\cdot)) \, d\mu$$

for all functions $f : A_1 \to \mathbb{R}$ of the form

$$f(\rho) = \prod_{a \in F} \rho(a)^{m(a)}$$

where the $m(a)$ are nonnegative integers. For a cube $\Lambda$ we let $\rho_\Lambda(\omega) = N(\omega_\Lambda)/|\Lambda|$. Then $\rho(\cdot) = \lim_{\Lambda \uparrow S} \rho_\Lambda(\cdot)$ $\mu$ - a. s. and $\mu_t$ - a. s. Furthermore, we have the estimate

$$\left| \frac{d}{ds} \int f(\rho_\Lambda(\cdot))\, d\mu_s \right| = \left| \int G(f \circ \rho_\Lambda)\, d\mu_s \right|$$

$$\leq \int \mu_s(d\omega) \sum_{\substack{x \in \Lambda \\ y \notin \Lambda}} c(xy, \omega) \sum_{a \in F} m(a) \left| \rho_\Lambda(a, {}^{xy}\omega) - \rho_\Lambda(a, \omega) \right|$$

$$\leq \left( K \sum_{a \in F} m(a) \right) |\Lambda|^{-1} \sum_{x \in \Lambda, y \notin \Lambda} c(y - x) = \varepsilon(\Lambda)$$

and therefore

$$\left| \int f(\rho_\Lambda(\ )) \, d\mu_t - \int f(\rho_\Lambda(\ ))\, d\mu \right| \leq \varepsilon(\Lambda)\, t .$$

Taking the limit $\Lambda \uparrow S$ the assertion follows since (3.36) implies $\lim_{\Lambda \uparrow S} \varepsilon(\Lambda) = 0$. ┘

The above remark shows that the function $t \to c(\mu_t)$ differs from the function

$$t \to f(\mu_t) = e(\mu_t) - h(\mu_t)$$

only by a constant. $f(\mu_t)$ is called the *specific free energy* of $\mu_t$ with respect to the potential $\Phi$. Because of (3.2) and (3.3) we have

$$f(\nu) = \lim_{\Lambda \uparrow S} f_\Lambda(\nu) / |\Lambda|$$

for any shift-invariant $\nu$, where

$$(3.38) \qquad f_\Lambda(\nu) = \int \nu(d\omega)\ [\ E_\Lambda(\ \omega_\Lambda|\alpha\ ) + \log \nu\ (\ X_\Lambda = \omega_\Lambda\ )\ ]$$

and $\alpha \in \Omega$ is an arbitrary fixed configuration. We will show that for each $\Lambda$ the function $t \to f_\Lambda(\mu_t)$ on $]\ 0, \infty\ [$ is differentiable. Up to a boundary term its derivative will be given by a function $d_\Lambda(\mu_t)$ which we are now going to define.

For $u, v \geq 0$ we put

$$h(u, v) = (u - v)\ (\log u - \log v) \quad ,$$

where we use the conventions $0 \log 0 = 0$ and $u \log 0 = -\infty\ (u > 0)$. $h(\cdot\ ,\cdot)$ is a lower semicontinuous function with values in $[0, \infty]$. For a probability measure $\nu$ we define

$$(3.39) \qquad d_\Lambda(\nu) = \frac{1}{2} \sum_{xy \subset \Lambda} \sum_{\zeta \in \Omega_\Lambda} h(\ T_\Lambda(\ xy, \zeta\ )\ ,\ T_\Lambda(\ xy, {}^{xy}\zeta\ )\ ) \ .$$

Here

$$(3.40) \qquad T_\Lambda(\ xy, \zeta\ ) = \int_{\{X_\Lambda = \zeta\}} c(\ xy, \cdot\ )\ d\nu \quad ,$$

and ${}^{xy}\zeta$ is defined by (2.1). The function $\nu \to d_\Lambda(\nu)$ takes values in $[0, \infty]$ and is lower semicontinuous.

As in (3.8) we let $\Lambda_n$ denote a cube of side $2^{n+1}$.

(3.41) <u>Remark:</u> *For any shift-invariant probability measure* $\nu$ *the sequence* $(\ d_{\Lambda_n}(\nu)/|\Lambda_n|\ )_{n\geq 1}$ *is increasing. Let* $d(\nu)$ *denote its limit in* $[0, \infty]$. *Then* $d(\cdot)$ *is lower semicontinuous and attains its minimum* $0$ *exactly on the set* $C_\theta$.

<u>Proof:</u> First we show that the sequence $d_{\Lambda_n}(\nu)/|\Lambda_n|$ $(n \geq 1)$ is increasing. (This will imply the existence and lower semicontinuity of $d(\nu)$.) We fix an integer $n \geq 1$ and let $V_1, \ldots, V_{2^d}$ denote pairwise disjoint translates of $\Lambda_n$ which

form a partition of $V = \Lambda_{n+1}$ . Then

$$2^d \; d_{\Lambda_n}(\nu) \;=\; \sum_{i=1}^{2^d} \; \sum_{xy \subset V_i} \; \sum_{\zeta_i \in \Omega_{V_i}} h(\, T_{V_i}(\, xy, \zeta_i \,) \, , \; T_{V_i}(\, xy, {}^{xy}\zeta_i \,)\,)$$

$$\leq \; \sum_{i=1}^{2^d} \; \sum_{xy \subset V_i} \; \sum_{\zeta \in \Omega_V} h(\, T_V(\, xy, \zeta \,) \, , \; T_V(\, xy, {}^{xy}\zeta \,)\,)$$

$$\leq \; d_{\Lambda_{n+1}}(\nu) \quad .$$

The second inequality follows from the nonnegativity of $h$ , and the first from the identity

$$T_{V_i}(\, xy, \zeta_i \,) \;=\; \sum_{\zeta \in \Omega_V : \zeta_{V_i} = \zeta_i} T_V(\, xy, \zeta \,)$$

and the subadditivity

$$h(\, \sum_1^k u_j \, , \; \sum_1^k v_j \,) \;\leq\; \sum_1^k \; h(\, u_j, v_j \,)$$

of $h$ .

Next we prove that $d(\nu) = 0$ iff $\nu \in C_\theta$ . Indeed, $d(\nu) = 0$ iff for arbitrarily large $\Lambda$ we have $d_\Lambda(\nu) = 0$ . This in turn is true iff for all $xy \subset \Lambda$ and $\zeta \in \Omega_\Lambda$ $T_\Lambda(\, xy, \zeta \,) = T_\Lambda(\, xy, {}^{xy}\zeta \,)$ and therefore iff equation (2.16) holds. By assumption the rate function $c(\cdot\, , \cdot)$ is irreducible. Thus (2.14) guarantees that (2.16) is equivalent to the assertion $\nu \in C_\theta$ . ┘

(3.42) Theorem: *Suppose that* $(P_t)_{t\geq 0}$ *is the transition semigroup of a particle jump process for a rate function satisfying conditions* (i) - (iv) . *Then for each shift-invariant probability measure* $\mu$ *on* $(\Omega, F)$ *and any* $t > 0$ *we have*

$$c(\, \mu P_t \,) \;\leq\; c(\mu) \;-\; \int_0^t d(\, \mu P_s \,)\, ds \quad .$$

Before proving this result we first state two corollaries. Combined with (3.23) the first one again yields the canonical variational principle (3.26), however this time under a somewhat stronger condition on $\Phi$.

(3.43) Corollary: *Assume*

$$\sum_{A \ni 0} ( |A| - 1 ) \| \Phi(A, \cdot) \| < \infty$$

*and* $\mu$ *is a shift-invariant probability measure on* $(\Omega, F)$. *Then* $c(\mu) = 0$ *implies* $\mu \in C$.

Proof: The condition on $\Phi$ ensures that there is a rate function $c(\cdot, \cdot)$ satisfying (i) - (iv) and the assumptions of the existence theorem (2.4), e.g. any of the functions (2.6) - (2.8). Now if $\mu \notin C$ then by (3.41) there is a $\delta > 0$ such that $\mu$ belongs to the open set $U = \{ \nu : d(\nu) > \delta \}$. Since $\mu_s \to \mu (s \to 0)$ we have $\mu_s \in U$ for sufficiently small $s > 0$. Thus (3.42) gives for sufficiently small $t > 0$

$$0 \leq c(\mu_t) \leq c(\mu) - t\delta < c(\mu) .$$

The second consequence of (3.42) is a statement on the asymptotic distribution of homogeneous particle jump process.

(3.44) Corollary: *Assume that the conditions of* (3.42) *are satisfied. Moreover, suppose that* $\nu$ *is the weak limit of a sequence* $(\mu P_{t_n})_{n \geq 1}$ *with* $\lim_{n \to \infty} t_n = \infty$. *Then* $\nu \in C_\theta$. *In particular, if* $\mu = \mu P_t$ *for all* $t$ *then* $\mu \in C_\theta$.

Proof: Assume $\nu \notin C_\theta$. Then $\nu \in U = \{ d(\cdot) > \delta \}$ for some $\delta > 0$. The mapping $P : (t, \mu) \to \mu_t$ is continuous; this follows immediately from the estimate

$$\left| \int f \, d( \mu_t - \mu'_{t+s} ) \right| \leq \left| \int P_t f \, d( \mu - \mu' ) \right| + \| P_s f - f \|$$

which is valid for all $t, s > 0$, $f \in C(\Omega)$ and probability measures $\mu, \mu'$.

Thus $P^{-1}(U)$ is an open neighbourhood of $(0, \nu)$, and we can find some $t > 0$ and an open neighbourhood $U'$ of $\nu$ such that $[0, t] \times U' \subset P^{-1}(U)$. For sufficiently large $n$ we have $\mu_{t_n} \in U'$ and therefore $d(\mu_{t_n+s}) > \delta$ for $0 \leq s \leq t$. From (3.42) we get $c(\mu_{t_n+t}) \leq c(\mu_{t_n}) - t\delta$ for all sufficiently large $n$. On the other hand, (3.42) shows that $s \to c(\mu_s)$ is non-increasing. Then

$$c(\mu) - \lim_{s \to \infty} c(\mu_s) = \infty$$

in contradiction to the estimates $c(\mu_s) \geq 0 (s \geq 0)$ and $c(\mu) \leq 2 \| \Phi \| + \log |F|$ which follow from (3.21), (3.2), (3.3), and (3.30). ┘

(3.45) Proof of Theorem (3.42):

In the definition (2.3) of the pregenerator $G$ of $(P_t)_{t \geq 0}$ we need to extend the sum only over those $xy \subset S$ for which $c(y - x) > 0$. This is a consequence of condition (ii). Therefore we will only consider such two-point sets in $S$ during the proof, without always explicity stating this restriction. Further, for the sake of brevity we will suppress the time index $s$ and write $\nu$ instead of $\mu_s$.
We fix a cube $\Lambda \in \mathcal{S}$.

1. For any $\zeta \in \Omega_\Lambda$ the function $s \to \nu(X_\Lambda = \zeta)$ on $]0, \infty[$ is twice continuously differentiable. Indeed, $f = 1_{\{X_\Lambda = \zeta\}}$ belongs to the domain of the generator $\overline{G}$ of $(P_t)_{t \geq 0}$. Thus the first derivative exists and is given by $\int Gf \, d\nu$. Moreover, the assumptions on the rate function in the existence theorem (2.4) ensure that $Gf$ also belongs to the domain of $\overline{G}$. This implies the assertion.

2. Suppose that $g \geq 0$ is a twice continuously differentiable function on $]0,\infty[$. Then the function $g \log g$ is differentiable with derivative $[1 + \log g] \frac{d}{ds} g$. Clearly this is true at every point $s$ with $g(s) > 0$. Therefore we may assume $g(s) = 0$. Then also $\frac{d}{ds} g(s) = 0$, and a second order approximation of $g$ shows that $(g(s') \log g(s')/(s' - s) \to 0$ as $s' \to s$, which is the desired result because of the convention $0 \log 0 = 0$.

3. Let $\zeta \in \Omega_\Lambda$, $xy \subset \Lambda$, and $s > 0$. Then $\nu(X_\Lambda = \zeta) > 0$ iff $\nu(X_\Lambda = {}^{xy}\zeta) > 0$. For if $\nu(X_\Lambda = \zeta) = 0$ then also

$$0 = \frac{d}{ds}\,\nu(X_\Lambda = \zeta) = \int \nu(d\omega) \sum_{uv\cap\Lambda\neq\emptyset} c(uv,\omega)\; 1_{\{X_\Lambda=\zeta\}}({}^{uv}\omega) .$$

Hence $\nu(X_\Lambda = {}^{uv}\zeta) = 0$ for all $uv \subset \Lambda$ with $c(u-v) > 0$.

4. From step 1. and 2. we see that the derivative

$$\frac{d}{ds} \sum_{\zeta\in\Omega_\Lambda} \nu(X_\Lambda = \zeta)\; \log \nu(X_\Lambda = \zeta)$$

of the entropy part of $f_\Lambda(\nu)$ exists and is given by

$$\sum_{\zeta\in\Omega_\Lambda} \log \nu(X_\Lambda = \zeta) \int G\, 1_{\{X_\Lambda=\zeta\}}\, d\nu ,$$

since

$$\sum_{\zeta\in\Omega_\Lambda} \frac{d}{ds}\,\nu(X_\Lambda = \zeta) = 0 .$$

More explicitly, the derivative is given by the expression

$$\int \nu(d\omega) \sum_{xy\cap\Lambda\neq\emptyset} c(xy,\omega)\; \log [\, \nu( X_\Lambda = ({}^{xy}\omega)_\Lambda ) \,/\, \nu( X_\Lambda = \omega_\Lambda ) \,] . \tag{3.47}$$

Observe that for $\nu$ - a. a. $\omega$ the expression in the logarithm in (3.47) is well-defined and finite. Indeed, for $\nu$ - a. a. $\omega$ we have for any $\Delta \supset \Lambda \cup xy$ $\nu(X_\Delta = \omega_\Delta) > 0$ and (by step 3. ) $\nu(X_\Delta = {}^{xy}\omega_\Delta) > 0$ and therefore $\nu(X_\Lambda = \omega_\Lambda) > 0$ and $\nu(X_\Lambda = {}^{xy}\omega_\Lambda) > 0$.

5. Since $E_\Lambda(X_\Lambda|\alpha)$ belongs to the domain of $G$, the energy part of $f_\Lambda(\nu)$ is also differentiable, and we have

$$\frac{d}{ds} \int \nu(d\omega)\;\; E_\Lambda(\, \omega_\Lambda | \alpha \,) = \tag{3.48}$$

$$= \int \nu(d\omega) \sum_{xy \cap \Lambda \neq \emptyset} c(xy, \omega) \; [\, E_\Lambda( {}^{xy}\omega_\Lambda | \alpha ) - E_\Lambda( \omega_\Lambda | \alpha ) \,] \quad .$$

6. The essential terms in (3.47) and (3.48) are those for which $xy \subset \Lambda$ , but first we consider the boundary terms with $x \in \Lambda$ , $y \notin \Lambda$ . In (3.48) these terms have the upper bound

$$K \; c(y - x) \; 2 \, \| \Phi \| \quad .$$

In (3.47) we get an upper bound if we replace log by $\log_+ = 0 \vee \log$ and $c(xy, \omega)$ by $K \; c(y - x)$ . This gives the expression

$$K \; c(y - x) \sum_{a \in F} \sum_{\zeta \in \Omega_\Lambda} \nu( X_{\Lambda \cup y} = \zeta a ) \; \log_+ [\, \nu( X_\Lambda = a\zeta_{\Lambda \setminus x} ) / \nu( X_\Lambda = \zeta ) \,]$$

Using the function $h'(u) = u \log_+ \frac{1}{u} \leq 1$ the double sum can be written as

$$\sum_{a,\zeta} \nu( X_y = a | X_\Lambda = \zeta ) \; \nu( X_\Lambda = a\zeta_{\Lambda \setminus x} ) \; h'( \nu(X_\Lambda = \zeta) / \nu(X_\Lambda = a\zeta_{\Lambda \setminus x}) )$$

and therefore is dominated by $|F|$ .

Introducing the notation

$$r_1(\Lambda) = ( 2 \| \Phi \| + |F| ) \; K \sum_{x \in \Lambda, y \notin \Lambda} c(y - x)$$

we have thus the result

$$\text{(3.49)} \qquad \frac{d}{ds} f_\Lambda(\nu) - r_1(\Lambda) \leq$$

$$\leq \sum_{xy \subset \Lambda} \sum_{\zeta \in \Omega_\Lambda} T_\Lambda(xy, \zeta) \; [ \log \nu( X_\Lambda = {}^{xy}\zeta ) + E_\Lambda( {}^{xy}\zeta | \alpha )$$

$$- \log \nu( X_\Lambda = \zeta ) + E_\Lambda(\zeta|\alpha) ] \quad .$$

Moreover, (3.36) gives

$$\lim_{\Lambda \uparrow S} r_1(\Lambda) / |\Lambda| = 0 \quad . \tag{3.50}$$

7. There is a number $r_2(\Lambda) < \infty$ such that for all $s > 0$

$$\frac{d}{ds} f_\Lambda(\nu) \leq r_2(\Lambda) - d_\Lambda(\nu) \quad . \tag{3.51}$$

Indeed, a simple change of variable shows that

$$- d_\Lambda(\nu) = \sum_{xy \subset \Lambda} \sum_{\zeta \in \Omega_\Lambda} T_\Lambda(xy, \zeta) \; [ \log T_\Lambda(xy, {}^{xy}\zeta) - \log T_\Lambda(xy, \zeta) ] \tag{3.52}$$

We thus compare the r. h. s. of (3.49) with that of (3.52) . In both expressions we need only to sum over those $\zeta$ for which $T_\Lambda(xy, \zeta) > 0$ and therefore $\nu(X_\Lambda = \zeta) > 0$ and (by step 3. ) also $\nu(X_\Lambda = {}^{xy}\zeta) > 0$ . For these $\zeta$ condition $(R_\Phi)$ gives the identity

$$E_\Lambda( {}^{xy}\zeta | \alpha ) - E_\Lambda( \zeta | \alpha ) =$$

$$= \log c( xy, {}^{xy}\zeta\alpha_{S \setminus \Lambda} ) - \log c( xy, \zeta\alpha_{S \setminus \Lambda} ) \quad ,$$

and we have the estimate

$$| \log [ \nu( X_\Lambda = \zeta ) \, c( xy, \zeta\alpha_{S \setminus \Lambda} ) ] - \log T_\Lambda( xy, \zeta ) |$$

$$\leq \frac{1}{c(y-x)} \int \nu(\, d\omega | X_\Lambda = \zeta\,) \;|\; c(\, xy,\, \omega_\Lambda\, \alpha_{S\setminus\Lambda}\,) - c(\, xy,\, \omega\,)\;|$$

$$\leq v_\Lambda(xy) \,/\, c(y-x) \quad ,$$

where

$$v_\Lambda(xy) = \max_{\zeta \in \Omega_\Lambda} \;|\; \max_{\omega:\omega_\Lambda=\zeta} c(\, xy,\, \omega\,) - \min_{\omega:\omega_\Lambda=\zeta} c(\, xy,\, \omega\,)\;| \quad . \tag{3.53}$$

Thus the r. h. s. of (3.49) is dominated by

$$- d_\Lambda(\nu) + \sum_{xy \subset \Lambda} \sum_{\zeta \in \Omega_\Lambda} T_\Lambda(\, xy,\, \zeta\,)\; 2\, v_\Lambda(xy) \,/\, c(y-x)$$

$$\leq - d_\Lambda(\nu) + 2\, K \sum_{xy \subset \Lambda} v_\Lambda(xy) \quad .$$

This proves our assertion if we put

$$r_2(\Lambda) = r_1(\Lambda) + 2\, K \sum_{xy \subset \Lambda} v_\Lambda(xy) \quad .$$

8. Next we show that $\lim_{\Lambda \uparrow S} r_2(\Lambda) / |\Lambda| = 0$ . Given $\varepsilon > 0$ we can find a cube $\Delta_1 \ni 0$ with

$$K \sum_{y \notin \Delta_1} c(y) < \varepsilon \quad .$$

Moreover, since for each $y \in S$ the function $c(y, \cdot\,)$ is uniformly continuous, there is a cube $\Delta_2 \supset \Delta_1$ such that for all $\Lambda \supset \Delta_2$

$$\sum_{y \in \Delta_1} v_\Lambda(0y) < \varepsilon \quad .$$

If $\Lambda$ is a sufficiently large cube then in addition we have

$$|\{ x \in \Lambda : x + \Delta_2 \not\subset \Lambda \}| / |\Lambda| < \varepsilon / \sum_y c(y) .$$

Combining these inequalities we obtain the estimate

$$\sum_{xy \subset \Lambda} v_\Lambda(xy) = \Big( \sum_{xy \subset \Lambda, y-x \notin \Delta_1} +$$

$$+ \sum_{\substack{xy \subset \Lambda \\ y-x\in\Delta_1, \Delta_2 \subset \Lambda - x}} + \sum_{\substack{xy \subset \Lambda \\ y-x\in\Delta_1, \Delta_2 \not\subset \Lambda - x}} \Big) v_{\Lambda-x} ( 0(y-x) )$$

$$\leq 3 |\Lambda| \varepsilon .$$

Because of (3.50) this completes the proof.

9. In steps 4. and 5. we have seen that the function $s \to f_\Lambda(\mu_s)$ is continuously differentiable on each interval $[\delta, t]$ with $0 < \delta < t$. Thus

$$f_\Lambda(\mu_t) - f_\Lambda(\mu_\delta) = \int_\delta^t \frac{d}{ds} f_\Lambda(\mu_s)\, ds .$$

Next we apply (3.51) and then let $\delta$ tend to zero. Because of the continuity of $f_\Lambda$ and the function $\delta \to \mu\delta$ we obtain

$$f_\Lambda(\mu_t) - f_\Lambda(\mu) \leq r_2(\Lambda)\, t - \int_0^t d_\Lambda(\mu_s)\, ds . \tag{3.54}$$

Dividing this inequality by $|\Lambda|$ and letting $\Lambda$ run through the sequence $(\Lambda_n)_{n\geq 1}$ we deduce from 8. and the monotone convergence theorem

$$f(\mu_t) - f(\mu) \leq \int_0^t d(\mu_s)\, ds .$$

Together with (3.37) this completes the proof of the theorem. $\lrcorner$

Notice that the proof above did not use the shift-invariance of $\mu$ until inequality (3.54). Since $d_\Lambda(\cdot) \geq 0$ we thus get the result that the function

$$t \to \limsup_{n \to \infty} f_{\Lambda_n}(\mu_t) \, / \, |\Lambda_n|$$

is non-increasing even when $\mu$ is not shift-invariant.

*Bibliographical notes*: Section 3.1 is based on H.O. Georgii (1975). Using different methods R.L. Thompson (1974) analysed a more general microcanonical model; his Theorem 3.1 contains Corollary (3.27) as a particular case. For a Markovian microcanonical model with a countable set $F$, the same assertion can already be found in D.A. Freedman (1962). The basic ideas of section 3.2 are due to R. Holley (1971), who considered a one-dimensional model with a particular rate function of the type (2.9). These ideas have been developed further for the so-called spin-flip process by several authors, the most recent being J. Moulin Ollagnier/ D. Pinchon (1977) (their paper contains also the earlier references). We have transferred some of their techniques back again to the particle jump process. In a more abstract setting W.G. Sullivan (1976) has proved various results relating to (3.44).

## § 4 Independent models

Here we consider systems of non-interacting particles with a spatially inhomogeneous self-potential. We will show that the extreme canonical Gibbs measures are Gibbs measures with respect to suitable activities if and only if the self-potential is homogeneous enough to ensure that a certain characteristic global quantity should be infinite.

### 4.1 The discrete case

We will confine ourselves to the classical situation in which each site $x$ of the countably infinite set $S$ is either occupied by a particle or remains empty. Thus we put $F = \{0, 1\}$ . As we mentioned in connection with definition (1.15) , we then may replace the set $A$ of activities by the interval $[0, \infty]$ . Moreover, every $L \in A_\Lambda$ is uniquely determined by the number $L(1)$ of particles. Thus we put $A_\Lambda = \{ 0,\dots,|\Lambda| \}$ , $N(\omega_\Lambda) = N(1, \omega_\Lambda)$ and

$$\Omega_{\Lambda,N} = \{ \omega \in \Omega_\Lambda : N(\omega_\Lambda) = N \}$$

( $0 \leq N \leq |\Lambda|$ ) . We fix a potential $\Phi$ such that $\Phi(A, \cdot)$ is constant whenever $|A| > 1$ . Normalizing $\Phi$ as in (1.14) we can write

$$(4.1) \qquad \Phi(A, \omega) = \begin{cases} 0 & \text{if } |A| > 1 \text{ or } A = \{x\}, \omega_x = 0 \\ -\log \sigma(x) & \text{otherwise} \end{cases}$$

for certain numbers $\sigma(x) > 0$ $(x \in S)$ . For such a $\Phi$ the Gibbs distributions $\gamma_\Lambda^z( \cdot \mid \omega )$ are independent of $\omega$ and equal to the product measure

$$(4.2) \qquad \gamma_\Lambda^z(\cdot) = \prod_{x\in\Lambda} \left( \frac{1}{1+z\sigma(x)} , \frac{z\sigma(x)}{1+z\sigma(x)} \right)$$

on $\Omega_\Lambda$. The canonical Gibbs distributions also do not depend on the boundary conditions $\omega$; for $0 \leq N \leq |\Lambda|$ and $\zeta \in \Omega_{\Lambda,N}$ they are given by

$$(4.3) \qquad \gamma_{\Lambda,N}(\zeta) = \gamma_\Lambda^z(\zeta \mid \Omega_{\Lambda,N}) = \prod_{x\in\Lambda} \sigma(x)^{\zeta_x} / Z_{\Lambda,N} .$$

Therefore we have the following characterization of the canonical Gibbs measures.

(4.4) Remark: *A probability measure* $\mu$ *on* $(\Omega, F)$ *is a canonical Gibbs measure with respect to the potential* (4.1) *iff for all* $\Lambda \in S$ *and* $\zeta \in \Omega_\Lambda$

$$\mu( X_\Lambda = \zeta \mid N(X_\Lambda) ) = \gamma_{\Lambda,N(X_\Lambda)}(\zeta) \qquad \mu\text{ - a. s.}$$

Proof: Since the $\sigma$ - algebra generated by $N(X_\Lambda)$ is contained in $E_\Lambda$, the condition is necessary. In order to prove its sufficiency we use the martingale convergence theorem. This shows that for $\mu$ - a. a. $\omega \in \{ N(X_\Lambda) = N(\zeta) \}$

$$\mu( X_\Lambda = \zeta \mid E_\Lambda )(\omega) =$$

$$= \lim_{\Delta \uparrow S\setminus\Lambda} \mu( X_\Lambda = \zeta , X_\Delta = \omega_\Delta ) / \mu( N(X_\Lambda) = N(\zeta) , X_\Delta = \omega_\Delta )$$

$$= \lim_{\Delta \uparrow S\setminus\Lambda} \gamma_{\Lambda\cup\Delta,N(\zeta\omega_\Delta)}(\zeta\omega_\Delta) / \sum_{\zeta':N(\zeta')=N(\zeta)} \gamma_{\Lambda\cup\Delta,N(\zeta\omega_\Delta)}(\zeta'\omega_\Delta)$$

$$= \gamma_{\Lambda,N(\zeta)}(\zeta) . \qquad \lrcorner$$

Next we will show that the set ex $C$ of extreme canonical Gibbs measures with respect to the potential (4.1) has a natural linear ordering. If $\mu$ and $\nu$ are probability measures on $(\Omega, F)$ we write $\mu \prec \nu$ when

$$(4.5) \qquad \mu( X_\Lambda \equiv 1 ) \leq \nu( X_\Lambda \equiv 1 ) \qquad (\Lambda \in S) \quad .$$

Observe that each probability measure $\mu$ is uniquely determined by its "correlation function" $\Lambda \to \mu( X_\Lambda \equiv 1 )$ ; this is a consequence of the Möbius inversion formula

$$(4.6) \qquad \mu( X_\Lambda = \zeta ) = \sum_{\zeta \subset \Delta \subset \Lambda} ( - 1 )^{|\Delta \setminus \zeta|} \mu( X_\Delta \equiv 1 ) \quad ,$$

where $\zeta \in \Omega_\Lambda$ is identified with the set $\{ x \in \Lambda : \zeta_x = 1 \}$ . This shows that $\mu = \nu$ whenever $\mu \prec \nu$ and $\nu \prec \mu$ . Clearly the Dirac-measures $\varepsilon_0$ and $\varepsilon_1$ (see (1.23) for the definition) are respectively the minimal and maximal elements in the ordering $\prec$ .

(4.7) Proposition: *In the ordering* $\prec$ *the set* ex $C$ *is linearly ordered.*

Proof: 1. Suppose that $\Delta \in S$ , $\emptyset \neq \Lambda \subset \Delta$ and $0 \leq N \leq M \leq |\Delta|$ . Then the correlation inequality

$$(4.8) \qquad \gamma_{\Delta,N} ( X_\Lambda = 1 ) \leq \gamma_{\Delta,M} ( X_\Lambda = 1 )$$

holds. This is evidently true if $|\Delta| = 1$ . For $|\Delta| > 1$ we use the following induction argument. It is sufficient to consider only the case when $M = N + 1$ , and this case is nontrivial only when $0 < N < |\Delta| - 1$ and $\Lambda \neq \Delta$ . Hence we assume that there is an $x \in \Delta \setminus \Lambda$ and define

$$p = \gamma_\Delta^1( X_x = 1 ) = 1 - q \quad ,$$

$$a_N = \gamma_{\Delta\setminus x}^1 ( X_\Lambda = 1 , N( X_{\Delta\setminus x} ) = N ) \quad ,$$

$$b_N = \gamma_{\Delta\setminus x}^1 ( N( X_{\Delta\setminus x} ) = N ) \quad .$$

Then from (4.3) we get

$$a_N / b_N = \gamma_{\Delta \setminus x, N} ( X_\Lambda = 1 ) ,$$

and the induction hypothesis ensures that

$$a_{N-1} / b_{N-1} \leq a_N / b_N \leq a_{N+1} / b_{N+1} .$$

Now we use the fact that $\gamma_\Delta^1$ is a product measure, and again (4.3) . Thus we conclude

$$\gamma_{\Delta,N} ( X_\Lambda = 1 ) = \frac{a_{N-1} p + a_N q}{b_{N-1} p + b_N q} \leq \frac{a_N}{b_N} \leq$$

$$\leq \frac{a_N p + a_{N+1} q}{b_N p + b_{N+1} q} = \gamma_{\Delta,N+1} ( X_\Lambda = 1 ) .$$

2. If $\mu, \nu \in \text{ex}\, C$ then $\mu \prec \nu$ or $\nu \prec \mu$ . We may assume that $\mu \neq \nu$ . Then we deduce from (4.6) that there is some $V \in S$ such that, say ,

$$\mu( X_V = 1 ) < \nu( X_V = 1 ) .$$

If $(\Delta_k)_{k\geq 1}$ is a sequence in $S$ increasing to $S$ then (1.32) (b) asserts that there are sequences $(N_k)_{k\geq 1}$ and $(M_k)_{k\geq 1}$ of integers such that for all $\Lambda \in S$

$$\mu( X_\Lambda = 1 ) = \lim_{k \to \infty} \gamma_{\Delta_k, N_k} ( X_\Lambda = 1 )$$

$$\nu( X_\Lambda = 1 ) = \lim_{k \to \infty} \gamma_{\Delta_k, M_k} ( X_\Lambda = 1 )$$

Choosing $\Lambda = V$ , we see from the correlation inequality above that for sufficiently large $k$ necessarily $N_K < M_k$ . This in turn implies (again by the correlation inequality) that for arbitrary $\Lambda$ $\mu( X_\Lambda = 1 ) \leq \nu( X_\Lambda = 1 )$ . This completes the proof . ⌋

Notice that the set $\bigcup_{0 \leq z \leq \infty} G(z)$ is also linearly ordered in the ordering $\prec$ . Indeed, (4.2) shows that

$$(4.9) \qquad G(z) = \{ \pi^z \} \qquad ( 0 \leq z \leq \infty ) ,$$

where we let $\pi^z$ denote the product measure

$$(4.10) \qquad \pi^z = \prod_{x \in S} \left( \frac{1}{1 + z\sigma(x)} , \frac{z\sigma(x)}{1 + z\sigma(x)} \right)$$

( $\pi^\infty = \varepsilon_1$ ) . Obviously $\pi^z \prec \pi^{z'}$ when $z \leq z'$ .

Now we ask for a condition on $\sigma$ (introduced in (4.1) ) which guarantees that all $\pi^z$ are extreme in $C$ . Because of (1.32) (b) , this can be true only if the $\pi^z$ are pairwise mutually singular. This, however, can be easily checked by means of a criterion of Kakutani (1948) .

(4.11) Remark: *The product measures* $\pi^z$ $( 0 < z < \infty )$ *are either pairwise equivalent or pairwise mutually singular. The second case occurs iff*

$$(4.12) \qquad \sum_{x \in S} \sigma(x) \, ( 1 + \sigma(x) )^{-2} = \infty ,$$

*and this condition is equivalent to*

$$(4.13) \qquad \sum_{x \in S} \sigma(x) \wedge 1/\sigma(x) = \infty .$$

Proof: Suppose that $0 < z < z' < \infty$ . Kakutani's criterion asserts that $\pi^z$ and $\pi^{z'}$ are mutually singular iff

$$\prod_{x \in S} \left( 1 + \sqrt{z z'} \, \sigma(x) \right) / \sqrt{( 1 + z\sigma(x) ) ( 1 + z'\sigma(x) )} = 0$$

or, equivalently ,

$$\sum_{x\in S} ( z + z' - 2\sqrt{z z'} )\, \sigma(x) \,/\, ( 1 + z\sigma(x) )\,( 1 + z'\sigma(x) ) = \infty .$$

This, of course, is equivalent to (4.12) . The final assertion follows from the inequality

$$( \sigma(x) \wedge 1/\sigma(x) )/4 \leq \sigma(x) ( 1 + \sigma(x) )^{-2} =$$

$$= ( \sqrt{\sigma(x)} + 1 / \sqrt{\sigma(x)} )^{-2} \leq \sigma(x) \wedge 1/\sigma(x) . \quad \lrcorner$$

We will prove below that condition (4.13) even implies that $\text{ex}\, C = \{ \pi^z : 0 \leq z \leq \infty \}$ . But first we want to consider the case when (4.13) is violated. Hence we assume

$$(4.14) \qquad \sum_{x\in S} \sigma(x) \wedge 1/\sigma(x) < \infty .$$

We then can find a partition $S = S_0 \cup S_1$ of $S$ such that $|S_i| = 0$ or $\infty$ $(i = 0, 1)$ and

$$(4.15) \qquad \sum_{x\in S_0} \sigma(x) < \infty , \qquad \sum_{x\in S_1} 1/\sigma(x) < \infty .$$

(Indeed, we can take an obvious modification of the partition $S = \{ \sigma(.) \leq 1 \} \cup \{ \sigma(.) > 1 \}$ . ) We let $I$ denote: the set of all integers if $|S_0| = |S_1| = \infty$ ; the set of all nonnegative integers if $S_1 = \emptyset$ ; and the set of all nonpositive integers if $S_0 = \emptyset$ . Observe that $I$ depends only on $\sigma$ and not on the particular choice of the partition. For $k \in I$ we define

$$\Omega_k = \{ \omega \in \Omega : N(1, \omega_{S_0}) + N(0, \omega_{S_1}) < \infty ,$$

$$N(1, \omega_{S_0}) - N(0, \omega_{S_1}) = k \} .$$

Clearly $\Omega_k \in E_\infty$ .

(4.16) Remark: *Assume that (4.14) holds. Then for all* $0 < z < \infty$ *the* $\sigma$ *- algebra* $E_\infty$ *is purely atomic with respect to* $\pi^z$ *with atoms* $\Omega_k$ , $k \in I$ .

Proof: Let $\Omega_c = \bigcup_{k\in I} \Omega_k$ . $\Omega_c$ is countable, and (4.15) gives $\pi^z(\Omega_c) = 1$ .
For each $\omega$ we have by definition

$$\pi^z(\{\omega\}) = \prod_{x\in S} (z\sigma(x))^{\omega_x} / (1 + z\sigma(x)) ,$$

and again (4.15) shows that this expression is strictly positive whenever $\omega \in \Omega_c$ . Thus $\pi^z(\Omega_k) > 0$ for all $k \in I$ . Finally, each configuration in $\Omega_k$ can be obtained from any other configuration in $\Omega_k$ by a permutation of finitely many coordinates; therefore $\Omega_k$ has no nontrivial $E_\infty$ - measurable subset. ┘

The above remark shows a second time that, under the condition (4.14) , the product measures $\pi^z$ ( $0 < z < \infty$ ) are pairwise equivalent and therefore not extreme in $C$ . On the other hand, it suggests that we consider the conditional probabilities $\pi^z(\cdot \mid \Omega_k)$ . These belong to $C$ since $\Omega_k \in E_\infty$ . Moreover, they do not depend on $z$ . This follows from

(4.17) Remark: *Let (4.14) be satisfied. Then for each* $k \in I$ *there is a unique* $\mu \in C$ *with* $\mu(\Omega_k) = 1$. $\mu$ *is given by*

$$(4.18) \qquad \mu(\{\omega\}) = c \prod_{x\in S_0} \sigma(x)^{\omega_x} \prod_{x\in S_1} \sigma(x)^{\omega_x - 1} \qquad (\omega \in \Omega_k) ,$$

*where* c *is a normalizing constant.*
(Observe that only finitely many factors in (4.18) are different from 1. )

Proof: Suppose that $\mu \in C$ and $\mu(\Omega_k) = 1$ . Then there is an $\omega \in \Omega_k$ such that $\mu(\{\omega\}) > 0$ . Any other $\zeta \in \Omega_k$ differs from $\omega$ only at finitely many sites. Therefore

$$\mu(\{\zeta\})\,/\,\mu(\{\omega\}) = \lim_{\Lambda\uparrow S} \mu(X_\Lambda = \zeta_\Lambda)\,/\,\mu(X_\Lambda = \omega_\Lambda)$$

$$= \lim_{\Lambda\uparrow S} \gamma_{\Lambda,N(\omega_\Lambda)}(\zeta_\Lambda)\,/\,\gamma_{\Lambda,N(\omega_\Lambda)}(\omega_\Lambda)$$

$$= \prod_{x:\zeta_x \neq \omega_x} \sigma(x)^{\zeta_x}\,/\,\sigma(x)^{\omega_x} = p(\zeta)\,/\,p(\omega) ,$$

where $p(\omega)$ denotes the r. h. s. of (4.18) . ┘

For $k \in I$ we let

(4.19) $$\nu^k = \pi^z(\,\cdot \mid \Omega_k\,)$$

denote the unique measure in $C$ which is concentrated on $\Omega_k$ . Moreover we put $\nu^{-\infty} = \varepsilon_0$ , $\nu^{\infty} = \varepsilon_1$ and let $I^*$ denote the compactification of $I$ by one or both of the points $+\infty$ and $-\infty$ . (If $I$ is bounded from below or above then already $\nu^0 = \varepsilon_0$ or $\varepsilon_1$ , respectively. ) Evidently, $\nu^k \prec \nu^{k'}$ iff $k \leq k'$ $(k, k' \in I^*)$ .

(4.20) <u>Theorem:</u> *Suppose* (4.14) *holds. Then*

$$\text{ex } C = \{\,\nu^k : k \in I^*\,\} .$$

<u>Proof:</u> Each $\nu^k$ is extreme. For if $\nu^k = \alpha\mu + (1-\alpha)\,\mu'$ with $0 < \alpha < 1$ and $\mu, \mu' \in C$ then both $\mu$ and $\mu'$ satisfy the condition of Remark (4.17) . Hence $\mu = \nu^k = \mu'$ . Conversely, we now suppose that $\mu \in \text{ex } C$ and show that $\mu = \nu^k$ for some $k \in I^*$ . (4.7) implies that either $\mu \prec \nu^k (\,k \in I\,)$ or $\nu^k \prec \mu (\,k \in I\,)$ or $\nu^k \prec \mu \prec \nu^{k+1}$ for some $k \in I$ . In the first case we have for all $\Lambda \in \mathcal{S}$ and $z > 0$

$$\mu( X_\Lambda = 1 ) \leq \sum_{k \in I} \pi^z(\Omega_k)\, \nu^k( X_\Lambda = 1 ) = \pi^z( X_\Lambda = 1 ) ;$$

in the limit $z \to 0$ we get $\mu = \varepsilon_0$ . In the second case a similar argument shows $\mu = \varepsilon_1$ . In the third case

$$\int N( 1, X_{S_0} )\, d\mu \leq \int N( 1, X_{S_0} )\, d\nu^{k+1} < \infty$$

$$\int N( 0, X_{S_1} )\, d\mu \leq \int N( 0, X_{S_1} )\, d\nu^{k} < \infty$$

and therefore $\mu( \bigcup_{k \in I} \Omega_k ) = 1$ . Since $\mu$ is trivial on $E_\infty$ there is an $l \in I$ with $\mu(\Omega_l) = 1$ , and from (4.17) we obtain $\mu = \nu^l$ . (Actually, $l = k$ or $k + 1$ ). $\lrcorner$

Now let us show that

$$\text{ex } C = \bigcup_z \text{ ex } G(z)$$

whenever

(4.21) $$\sum_{x \in S} \sigma(x) \wedge 1/\sigma(x) = \infty .$$

(4.22) Theorem: *Suppose (4.21) is satisfied. Then*

$$\text{ex } C = \{ \pi^z : 0 \leq z \leq \infty \}$$

Proof: We will prove in Theorem (4.23) below that for any $0 < z < \infty$ the product measure $\pi^z$ is trivial on $E_\infty$ and therefore extreme in $C$ whenever the condition

$$\sum_{x\in S} z\sigma(x) \; ( 1 + z\sigma(x) )^{-2} = \infty \; ,$$

which is equivalent to (4.21) , holds. Thus here we must only show that $C$ has no other extreme points. The argument is similar to the proof of the corresponding statement in (4.20) . Let $\mu \in \text{ex } C$ .

We choose an $x \in S$ and a $\hat{z} \in [0, \infty]$ such that

$$\mu( X_x = 1 ) = \hat{z}\, \sigma(x) \,/\, ( 1 + \hat{z}\, \sigma(x) ) = \pi^{\hat{z}}( X_x = 1 ) \; .$$

Then for all $z > \hat{z}$ we have $\mu( X_x = 1 ) < \pi^z( X_x = 1 )$ and thus $\mu \prec \pi^z$ . Similarly, $\pi^z \prec \mu$ whenever $z < \hat{z}$ . Therefore

$$\sup_{z<\hat{z}} \pi^z( X_\Lambda = 1 ) \le \mu( X_\Lambda = 1 ) \le \inf_{z>\hat{z}} \pi^z( X_\Lambda = 1 )$$

for all $\Lambda \in S$ . This yields

$$\mu( X_\Lambda = 1 ) = \pi^{\hat{z}} ( X_\Lambda = 1 ) \qquad (\Lambda \in S)$$

and thus $\mu = \pi^{\hat{z}}$ . ┘

The next theorem (which completes the proof of (4.22) ) is a generalization to inhomogeneous product measures of the well-known Hewitt/Savage 0 - 1 law .

(4.23) <u>Theorem:</u> *Suppose that* $\pi$ *is a product probability measure on* $(\Omega, F)$ , *say*

$$\pi = \prod_{x\in S} ( q(x), p(x) )$$

*with* $p(x) = \pi( X_x = 1 ) = 1 - q(x)$ . *Then* $\pi$ *is trivial on* $E_\infty$ *iff*

$$\sum_{x \in S} p(x)\, q(x) = 0 \quad \text{or} \quad \infty \quad .$$

Proof: 1. "only if" : In the case $0 < p(x) < 1$ $(x \in S)$ this has already been seen in (4.16) . It is easy to extend that proof to the general case $0 \leq p(x) \leq 1$ $(x \in S)$ , but the following argument is more elegant: The random variables $X_x - p(x)$ $(x \in S)$ on $(\Omega, F, \pi)$ are independent and have expectation 0 and variance $p(x)\, q(x)$ . If $0 < \sum_{x \in S} p(x)\, q(x) < \infty$ then $Y = \sum_{x \in S} (X_x - p(x))$ $\pi$ - a. s. exists, is $E_\infty$ - measurable, and is not a. s. constant since its variance is positive. Thus $\pi$ cannot be trivial on $E_\infty$ .

2. "if" : It is convenient to identify $S$ and $\{1, 2, \dots\}$ . For each $k \geq 1$ we put $N_k = N( X_{\{1,\dots,k\}} ) = \sum_1^k X_i$ . We fix a set $A \in E_\infty$ . For each $k \geq 1$ , $0 \leq n \leq k$ there is a set $B_{k,n} \in F_{\{k+1,k+2,\dots\}}$ such that

$$A \cap \{ N_k = n \} = B_{k,n} \cap \{ N_k = n \} \quad .$$

Defining $f_k(n) = \pi( B_{k,n} )$ we obtain the formula

$$\pi( A \mid F_{\{1,\dots,k\}} ) = \sum_{n=0}^{k} 1_{\{N_k = n\}}\, \pi( B_{k,n} ) = f_k( N_k ) \quad ,$$

and the martingale convergence theorem implies

$$\text{(4.24)} \qquad A = \{ \lim_{k \to \infty} f_k( N_k ) = 1 \} \qquad \pi \text{ - a. s.}$$

(Notice that this is the well-known construction relating the tail field of the random walk $( N_k )_{k \geq 1}$ with its space-time harmonic functions.) In order to prove the equation $\pi(A) = \pi(A)^2$ we introduce the product measure $\mu = \pi \otimes \pi$ on $\Omega \times \Omega$ . Furthermore, we define on $\Omega \times \Omega$ the projections $X_i^1 (\omega, \zeta) = \omega_i$, $X_i^2 (\omega, \zeta) = \zeta_i$ $(i \geq 1)$ and the counting variables $N_k^1 = \sum_1^k X_i^1$ , $N_k^2 = \sum_1^k X_i^2$ $(k \geq 1)$ . Then

(4.24) gives

$$\pi(A)^2 = \mu( \lim_{k \to \infty} f_k( N_k^1 ) = 1 , \lim_{k \to \infty} f_k( N_k^2 ) = 1 ) .$$

If we could prove that

$$(4.25) \qquad \mu( N_k^1 = N_k^2 \text{ for infinitely many } k ) = 1$$

then this would imply that the events $\{ \lim_{k \to \infty} f_k( N_k^j ) = 1 \}$ $(j = 1, 2)$ $\mu$ - a.s coincide (Remember that the limits $\mu$ - a. s. exist) . Consequently we would have

$$\pi(A)^2 = \mu( \lim_{k \to \infty} f_k( N_k^1 ) = 1 ) = \pi(A)$$

and therefore $\pi(A) = 0$ or $1$ . Thus let us verify (4.25) . Consider the sequence

$$M_k = N_k^1 - N_k^2 = \sum_1^k ( X_i^1 - X_i^2 ) =: \sum_1^k Y_i .$$

The random variables $(Y_i)_{i \geq 1}$ are independent with respect to $\mu$ , and their distribution is given by

$$\mu( Y_i = 1 ) = \mu( Y_i = - 1 ) = p(i)\, q(i)$$

$$\mu( Y_i = 0 ) = 1 - 2\, p(i)\, q(i) .$$

Therefore if $\Sigma\, p(i)\, q(i) = 0$ then $Y_i = 0$ $\mu$ - a. s. for all $i$ and thus $M_k = 0$ $\mu$ - a. s. for all $k$ . If $\Sigma\, p(i)\, q(i) = \infty$ then the Borel-Cantelli-Lemma asserts that

$$\mu( Y_i = \pm 1 \text{ i. o. } ) = 1 ;$$

thus $(M_k)_{k\geq 1}$ is a martingale with bounded increments which a. s. does not converge. This implies that

$$\limsup_{k\to\infty} M_k = \infty , \quad \liminf_{k\to\infty} M_k = -\infty \qquad \mu - \text{a. s.}$$

(We can also argue as follows: Up to random holding times, $(M_k)_{k\geq 1}$ is a symmetric simple random walk on the integers. Since $(M_k)_{k\geq 1}$ has infinitely many jumps it is recurrent.) Both arguments give (4.25) . ⌋

We conclude this section with a remark showing that the distribution of activities in the integral representation of canonical Gibbs measures by Gibbs measures comes from an "individual activity function" .

(4.26) Remark: *Suppose* (4.21) *holds and* $\mu \in C$ . *Then for* $\mu$ - a. a. $\omega$ *the limit*

$$z(\omega) = \lim_{\Lambda\uparrow S} Z_{\Lambda,N(\omega_\Lambda)-1} / Z_{\Lambda,N(\omega_\Lambda)}$$

*exists, and* $\mu$ *has the representation*

$$\mu = \int \pi^{z(\omega)} \mu(d\omega) .$$

Proof: We enlarge $S$ by an additional site $0$ and define $\underline{S} = S \cup \{0\}$ . We let $\sigma(0) = 1$ and consider the corresponding product measures $\underline{\pi}^z$ on $\underline{\Omega} = \{0, 1\}^{\underline{S}}$. (4.22) gives $\underline{\pi}^z \in \text{ex}\,\underline{C}$ , where $\underline{C}$ is the set of all canonical Gibbs measures w. r. to $\sigma$ on $\underline{\Omega}$ . For $\omega \in \Omega$ we put $\underline{\omega} = 0\,\omega \in \underline{\Omega}$ . Then the martingale convergence theorem shows that for $\underline{\pi}^z$ - a. a. $\underline{\omega}$ and thus for $\pi^z$ - a. a. $\omega$

$$z = \underline{\pi}^z ( X_0 = 1 ) / \underline{\pi}^z ( X_0 = 0 )$$

$$= \lim_{\Lambda\uparrow \underline{S}} \gamma_{\Lambda,N(\underline{\omega}_\Lambda)} ( X_0 = 1 ) / \gamma_{\Lambda,N(\underline{\omega}_\Lambda)} ( X_0 = 0 )$$

$$= \lim_{\Lambda \uparrow S} Z_{\Lambda, N(\omega_\Lambda)-1} \, / \, Z_{\Lambda, N(\omega_\Lambda)} \quad .$$

This completes the proof for $\mu \in \text{ex}\, C$ . For general $\mu \in C$ we obtain the result by using the representation of $\mu$ by extreme points. ⌋

Observe that the definition of $z(\omega)$ as a limit of ratios of canonical partition functions is consistent with formula (3.17) . As we will see in §§ 5 and 6 , this method of identifying the activity distribution in the representation of a canonical Gibbs measures also works in the case of a general interaction potential.

In the particular case when $\sigma(x) = 1 \quad (x \in S)$ the activity function $z(\omega)$ is a simple function of the particle density

$$\rho(\omega) = \lim_{\Lambda \uparrow S} N(\omega_\Lambda) \, / \, |\Lambda| \quad ,$$

namely

$$z(\omega) = \rho(\omega) \, / \, ( 1 - \rho(\omega) ) \quad .$$

In the case of a general $\sigma$ satisfying (4.21) we can find a similar expression for $z(\omega)$ using the strong law of large numbers; for details we refer the reader to p. 352 of Liggett (1976) .

## 4.2 The continuous case

In this section there are no difficulties in considering a setup more general than that which was introduced in section 2.2: We let $S$ denote a locally compact space with a countable base and $\mathcal{B}$ the $\sigma$ - algebra of Borel sets in $S$ . The configuration space will be $M$ , the set of all Radon measures on $S$ taking values only in the set $\{ 0, 1, 2, \ldots, \infty \}$ ; $M$ is endowed with the $\sigma$ - algebra $\mathcal{F}$ generated by the vague topology. We fix a (not necessarily diffuse) positive Radon measure

$\sigma \neq 0$ on $S$, and for each $z \geq 0$ we let $\pi^z$ denote the Poisson point process on $S$ with intensity measure $z\,\sigma$; in particular, we put $\pi = \pi^1$. Gibbs measures and canonical Gibbs measures are defined as in (1.63) and (1.68) for the particular case $\Phi \equiv 0$, except for the requirement, which we drop now, that they should be concentrated on $\Omega_0$. Then $G(z) = \{\pi^z\}$, $z \geq 0$, and $C$ has the following characterization similar to (4.4):

(4.27) Remark: *Suppose that* $\mu$ *is a probability measure on* $(M, F)$. *Then* $\mu \in C$ *iff for all compact sets* $\Lambda$ *and* $\mu$ - a. a. $\omega$ *the equation*

$$\mu(\,\cdot\mid N(\Lambda) = \omega(\Lambda)\,) \quad = \quad \pi(\,\cdot\mid N(\Lambda) = \omega(\Lambda)\,)$$

*holds.*

Proof: Similar to (4.4). ┘

A probability measure $\mu$ on $(M, F)$ is called a *mixed Poisson process* corresponding to $\sigma$ if $\mu$ has a representation

$$\mu \quad = \quad \int \pi^z \quad m(dz)$$

with a probability measure $m$ on $[0, \infty[$. Clearly each mixed Poisson process belongs to $C$. The converse question of whether each $\mu \in C$ is a mixed Poisson process is equivalent to our basic problem

(4.28) $$\text{ex } C \quad = \quad \bigcup_{z \geq 0} \text{ ex } G(z) \quad = \quad \{\, \pi^z \;:\; z \geq 0 \,\} \quad .$$

We will prove that (4.28) holds if and only if $\sigma(S) = \infty$. Notice that this condition is simpler than the corresponding condition in the discrete model. The reason is that, in contrast to the discrete model, $M$ does not contain a maximal configuration. Therefore in the present case the total particle number is the only symmetric variable which has to be prevented from being a.s. finite by a condition on $\sigma$.

(4.29) Theorem: *Assume* $\sigma(S) = \infty$ . *Then*

$$\text{ex } C = \{ \pi^z : z \geq 0 \} .$$

*Moreover, if* $\mu \in C$ *then for* $\mu$ *- a. a.* $\omega$ *the limit*

$$\rho(\omega) = \lim_{\Lambda \uparrow S} \omega(\Lambda) / \sigma(\Lambda)$$

*exists, and* $\mu$ *has the representation*

$$\mu = \int \pi^{\rho(\omega)} \mu(d\omega) .$$

Notice that the particle density $\rho(\omega)$ which governs the activity distribution of canonical Gibbs measures in this theorem has the same form as the function $z(\omega)$ in (4.26) . Indeed, since

$$Z_{\Lambda,N} = \pi( N(\Lambda) = N ) = e^{-\sigma(\Lambda)} \sigma(\Lambda)^N / N!$$

we have

$$\omega(\Lambda) / \sigma(\Lambda) = Z_{\Lambda,\omega(\Lambda)-1} / Z_{\Lambda,\omega(\Lambda)} .$$

Proof: Let $\mu \in \text{ex } C$ . We consider the Laplace transform $\Psi$ of $\mu$ defined by

$$\Psi(f) = \int \mu(d\omega) \exp [ - \int f \, d\omega ] ,$$

where $f$ is a nonnegative, bounded, measurable function on $S$ with compact support. The approximation theorem (1.74) asserts that for $\mu$ - a. a. $\omega$

$$(4.30) \qquad \Psi(f) = \lim_{\Lambda \uparrow S} \int \pi( d\zeta \mid N(\Lambda) = \omega(\Lambda) ) \exp [ - \int f \, d\zeta ]$$

$$= \lim_{\Lambda \uparrow S} \left[ \int_\Lambda e^{-f} \, d\sigma / \sigma(\Lambda) \right]^{\omega(\Lambda)}$$

$$= \lim_{\Lambda \uparrow S} \left[ 1 + m(f) / \sigma(\Lambda) \right]^{\omega(\Lambda)}$$

where

$$m(f) = \int ( e^{-f} - 1 ) \, d\sigma \quad .$$

Since $\sigma$ is infinite we know that

$$\lim_{\Lambda \uparrow S} \left[ 1 + m(f) / \sigma(\Lambda) \right]^{\sigma(\Lambda)} = \exp \, m(f) \quad .$$

Choosing a particular $f$ with $m(f) \neq 0$ we deduce from (4.30) that for $\mu$ - a.a.$\omega$ the limit $\rho(\omega)$ exists and does not depend on $\omega$, say $\rho(\omega) = \rho$. Moreover, for all $f$ we have

$$\Psi(f) = \exp \, \rho \, m(f) \quad .$$

Thus $\Psi$ coincides with the Laplace transform of $\pi^\rho$, showing that $\mu = \pi^\rho$. Thus we have proved that

$$\text{ex } C \subset \{ \pi^z : z \geq 0 \} \; ;$$

in particular we have seen that for any $\mu \in \text{ex } C$ $\mu( \rho(\cdot) \text{ exists } ) = 1$ and

$$\mu = \int_{\{\pi^{\rho(\cdot)} \in \text{ ex } C\}} \pi^{\rho(\omega)} \, \mu(d\omega) \quad .$$

By the usual integral representation argument these assertions extend to all $\mu \in C$. From this we deduce that $\pi^z \in \text{ex } C$ for all $z \geq 0$; indeed, the tail-measurable function $\rho( \; )$ is $\pi^z$ - a. s. constant. ┘

Now we consider the case when $\sigma$ is finite. Then for any $k \geq 0$ there is a measure $\nu^k \in C$ carried by the symmetric event $\{ N(S) = k \}$, namely

$$\nu^k = \pi( \cdot \mid N(S) = k ) .$$

Obviously, we have for any measurable function $f \geq 0$

(4.31) $$\int f \, d\nu^k = \sigma(S)^{-k} \int_{S^k} \sigma(dx_1) \ldots \sigma(dx_k) \quad f( \varepsilon_{x_1} + \ldots + \varepsilon_{x_2} ) .$$

(4.32) Theorem: *Suppose* $\sigma(S) < \infty$ . *Then*

$$\text{ex } C = \{ \nu^k : k \geq 0 \} .$$

*Every* $\mu \in C$ *is carried by* $\{ N(S) < \infty \}$ *and has the representation*

(4.33) $$\mu = \int \nu^{\omega(S)} \mu(d\omega) .$$

*The* $\sigma$ - *algebra* $E_\infty$ *is purely atomic with respect to any of the Poisson processes* $\pi^z$ $(z > 0)$ *with atoms* $\{ N(S) = k \}$ $(k \geq 0)$ .

Proof: Except for the final sentence, all statements can be verified in the same way as the corresponding statements in (4.29) ; notice that $\nu^k$ has the Laplace transform

$$[ 1 + m(f) / \sigma(S) ]^k .$$

In order to see that $E_\infty$ is atomic with respect to $\pi^z$ $(z > 0)$ we first observe that

$$\pi^z( N(S) = k ) = e^{-z\sigma(S)} z^k \sigma(S)^k / k! > 0 .$$

Now we assume that there are disjoint sets $A, B \in E_\infty$ such that $A \cup B \subset \{ N(S) = k \}$ for some $k$ and $\pi^Z(A)\, \pi^Z(B) > 0$ . Then $\pi^Z( \cdot \mid A )$ and $\pi^Z( \cdot \mid B )$ are different elements of $C$ which are concentrated on $\{N(S) = k\}$ . But (4.33) shows that this is impossible. $\lrcorner$

We conclude this section with a point-process counterpart to the Hewitt/Savage 0 - 1 law:

(4.34) <u>Corollary:</u> *Suppose that* $\pi$ *is a Poisson process with intensity measure* $\sigma$ . *Then* $\pi(A) = 0$ *or* 1 *for all* $A \in E_\infty$ *if and only if* $\sigma$ *is infinite.*

*Bibliographical notes*: The results of section 4.1 are due to Th.M. Liggett (1976) who studied the invariant measures for particle jump processes corresponding to rate functions $c( \cdot , \cdot )$ of the form (2.9) with $\Phi \equiv 0$ and jump matrices $p$ satisfying the reversibility condition (2.13) . (Thus $c( \cdot , \cdot )$ satisfies $(R_\Phi)$ with $\Phi$ as in (4.1) , and because of (2.14) Liggett's results imply those in section 4.1 . ) The statements (4.7), (4.22) , and (4.23) were independently proved by J. Pitman (1978) , (4.23) also by H. Rost (unpublished). Their ideas are similar to those used in this text which are based on an unpublished note of C. Preston. An extension of Theorem (4.23) to product measures on an infinite product of sets of cardinality larger than 2 can be found in D. Aldous/ J. Pitman (1977). The idea used in the proof of (4.29) is due to W. von Waldenfels; Nguyen Xuan Xanh and H. Zessin (1977) observed that it can be used to obtain (4.29). Different proofs are contained in J. Kerstan/K. Matthes/J. Mecke (1974) (see 1.6.10 and 1.6.12 ) and J. Neveu (1977) (p. 270) . In the case $S = \mathbb{R}$ , $\sigma = \lambda$ Theorem (4.29) was already proved by K. Nawrotzki (1962) and D.A. Freedman (1963) .

## § 5 Discrete models with interaction

We let again $S$ denote a countably infinite set of sites, $F$ a finite set of types of particles, and $\Omega = F^S$. We fix a potential $\Phi$ and consider the set $C = C(\Phi)$. We want to find general conditions on $\Phi$ which imply

$$\text{(5.1)} \qquad \text{ex } C = \bigcup_z \text{ ex } G(z) \quad .$$

In section 4.1 we have seen that for potentials of the form (4.1) the equation (5.1) holds iff the series $\Sigma\, \sigma(x) \wedge 1/\sigma(x)$ diverges. The first attempt to extend this criterion to general $\Phi$ is to consider the series $\Sigma\, m_x$ where

$$\text{(5.2)} \qquad m_x = \min_{\omega\in\Omega} \min_{a,b\in F} \exp[\, E_x(a|\omega) - E_x(b|\omega)\, ] \quad .$$

Since we take the minimum over $\omega$ we can only hope that the condition

$$\text{(5.3)} \qquad \sum_{x\in S} m_x = \infty$$

is sufficient for (5.1). Unfortunately, we are able to prove (5.1) only under the stronger conditions (A) and (B) below.

### 5.1 Formulation of results

The basic idea in the proof of (5.1) is the following. In order to verify that a measure $\mu \in \text{ex } C$ is a Gibbs measure with respect to an activity $z$ it is sufficient to consider the ratios

$$\text{(5.4)} \qquad \mu_x(a|\omega) \,/\, \mu_x(b|\omega)$$

of conditional probabilities. These can be obtained by two limiting procedures: (5.4) is the limit of

$$\mu( X_x = a \mid X_{W\setminus x} = \omega_{W\setminus x} ) \,/\, \mu( X_x = b \mid X_{W\setminus x} = \omega_{W\setminus x} )$$

as $W \uparrow S$, and (1.32) asserts that these ratios can be expressed in terms of canonical distributions in $\Lambda$ for $\Lambda \uparrow S$. The approximation term for (5.4) thus obtained consists of an energy part which is easy to handle and of the ratio

(5.5) $$z_{\Lambda,W,V} (a, b, \omega) =$$

$$= Z_{\Lambda\setminus W, N(\omega_{\Lambda\setminus W}) + 1_b - 1_a} ( \alpha_V \, \omega_{S\setminus V} ) \,/\, Z_{\Lambda\setminus W, N(\omega_{\Lambda\setminus W})} ( \alpha_V \, \omega_{S\setminus V} )$$

of canonical partition functions. Here $V \subset W \subset \Lambda \in \mathcal{S}$, $a, b \in F$, $\omega \in \Omega$; $\alpha \in \Omega$ is an arbitrarily fixed reference configuration, and $1_a$ denotes the function on $F$ which is equal to $1$ on $\{a\}$ and $0$ otherwise. The limiting procedures above lead to the function

(5.6) $$z( a, b, \omega ) = \limsup_{V\uparrow S} \; \limsup_{W\uparrow S} \; \limsup_{\Lambda\uparrow S} z_{\Lambda,W,V} ( a, b, \omega )$$

on $F \times F \times \Omega$ with values in $[0, \infty]$. (Under certain circumstances we will require that $\Lambda$ runs through a particular sequence depending on $\Phi$.) In section 5.2 we will see that

$$z_V( a, b, \cdot ) = \limsup_{W\uparrow S} \; \limsup_{\Lambda\uparrow S} z_{\Lambda,W,V} ( a, b, \omega )$$

a. s. does not depend on $V$. Thus we perform the limit $V \uparrow S$ only for technical reasons: This limit implies that the functions $z( a, b, \cdot )$ are measurable with respect to the tail field, and this will be important later on. Actually we will see that

$$z(a, b, \cdot) = \lim_{\Lambda \uparrow S} z_{\Lambda,\emptyset,\emptyset}(a, b, \cdot) \qquad \mu\text{ - a. s.}$$

for all $\mu \in C$ whenever (5.1) holds. Moreover, in section 7.1 we will show that for certain homogeneous models $z(a, b, \cdot)$ is a function of the particle density $\rho(\cdot)$. Here we will use the function $z(\cdot, \cdot, \cdot)$ in order to decide whether for a given $\mu \in \text{ex } C$ there is an activity $z$ such that $\mu \in G(z)$. The precise state of affairs is described in Proposition (5.9) below.

For any $\mu \in C$ we let

$$(5.7) \qquad F_\mu = \{ a \in F : \mu( N(a, \cdot) > 0 ) > 0 \}$$
$$= \{ a \in F : \mu( X_x = a ) > 0 \text{ for some } x \in S \}$$

denote the set of types of particles which occur in the typical configurations of $\mu$. In particular, for $\mu \in G(z)$ we have

$$F_\mu = \{ a \in F : z(a) > 0 \} .$$

(5.8) <u>Remark:</u> *Let* $\mu \in C$, $a \in F_\mu$. *Then* $\mu( X_x = a ) > 0$ *for all* $x \in S$.

<u>Proof:</u> Assume that $\mu( X_x = a ) = 0$ for some $x \in S$. By assumption there is an $y \in S$ with $\mu( X_y = a ) > 0$ and therefore some $b \neq a$ with $\mu( X_x = b, X_y = a ) > 0$ and thus $\mu( N(X_\Lambda) = L ) > 0$, where $\Lambda = \{x, y\}$ and $L = 1_a + 1_b$. This gives the contradiction

$$0 = \mu( X_x = a , X_y = b ) = \int_{\{N(X_\Lambda)=L\}} \gamma_{\Lambda,L}(ab \mid \cdot) \, d\mu > 0 . \qquad \lrcorner$$

We introduce the following notation: If $f$ is a measurable function on $\Omega$ which is a. s. constant with respect to a probability measure $\mu$ on $\Omega$ then we let $f(\mu)$ denote this constant value of $f$. In particular this notation applies if $f$ is symmetric or tail-measurable and $\mu$ is an extreme canonical Gibbs measure.

The next proposition will be proved in section 5.2 .

(5.9) Proposition: *Suppose* $\mu \in \text{ex } C$ . *Then the following statements are equivalent.*

(I) $\mu \in G(z)$ *for some* $z \in A$ .

(II) *For all* $x \in S$

$$\int d\mu \prod_{a \in F_\mu} \mu_x(a|\cdot) > 0 .$$

(III) *For any* $a, b \in F_\mu$ *there is some* $x \in S$ *such that*

$$\mu( X_x = a , \mu_x(b|\cdot) > 0 ) > 0 .$$

(IV) *For any* $a, b \in F_\mu$ *we have*

$$z( a, b, \mu ) > 0 .$$

(V) *For any* $a, b \in F_\mu$

$$z( a, b, \mu ) \vee z( b, a, \mu ) > 0 .$$

*In this case we have*

$$F_\mu = \{ a \in F : \mu( N(a,\cdot) = \infty ) = 1 \} .$$

*Moreover, if* $\mu \in G(z)$ *then* $z( a, b, \mu ) = z(a)/z(b)$ *for all* $a \in F$ , $b \in F_\mu$ . *Conversely,* (V) *implies*

$$z( a, b, \mu ) < \infty \qquad ( a \in F , b \in F_\mu )$$

*and*

$$z(a, b, \mu)\, z(b, c, \mu) = z(a, c, \mu) \qquad (a \in F,\ b, c \in F_\mu)$$

*and finally*

$$\mu \in G(z(\cdot, b, \mu)) \qquad (b \in F_\mu) \quad .$$

The equivalence (I) $\Longleftrightarrow$ (II) expresses some sort of "randomness principle" : An extreme canonical Gibbs measure is a Gibbs measure iff for each $x \in S$ it is impossible to determine the type of the particle at $x$ from the knowledge of the configuration on $S \setminus x$ (except when $\mu$ admits only one type of particle, i. e. $|F_\mu| = 1$ ). Statement (III) is equivalent to the assertion

$$(5.10) \qquad \sup_{x\in S} \int d\mu \;\; \mu_x(a|\cdot)\ \mu_x(b|\cdot) > 0 \qquad (a, b \in F_\mu)$$

and shows that it is sufficient to verify this randomness property of $\mu$ at only one site. This property is trivially satisfied by the unit masses $\varepsilon_a$ $(a \in F)$ which are characterized by the condition $|F_\mu| = 1$ .

Moreover, (III) implies

$$(5.11) \qquad \mu(N(a, \cdot) = \infty) = 1 \qquad (a \in F_\mu)$$

Indeed, if $\mu(N(a,\cdot) = \infty) < 1$ for some $a \in F_\mu$ then $|F_\mu| \geq 2$ and (due to the triviality of $E_\infty$ ) there is some $k \in \mathbb{N}$ with $\mu(N(a,\cdot) = k) = 1$ . Thus for all $x \in S$ and $\mu$ - a. a. $\omega$ we have $\mu_x(a|\omega) = 1$ or $0$ according to whether $N(a, \omega_{S\setminus x}) = k - 1$ or not. In particular, $\mu_x(a|\cdot) = 0$ a. s. on $\{X_x = b\}$ for all $b \in F_\mu \setminus \{a\}$ .

A refinement of this argument gives us a condition on $\Phi$ which is necessary for (5.1) . This condition reduces to the sufficient condition (4.21) whenever $|F| = 2$ and $\Phi$ is of the form (4.1) .

For $A, B \subset F$ we define

$$M_x(A, B) = \max_{\omega \in (A \cup B)^S} \sum_{a \in A} \exp[-E_x(a|\omega)] / \sum_{b \in B} \exp[-E_x(b|\omega)] .$$

(5.12) Corollary: *In order that the relation*

$$\text{ex } C \supset \bigcup_{z \in A} \text{ex } G(z)$$

*should hold it is necessary that for any two disjoint* $A, B \subset F$

$$\sum_{x \in S} M_x(A, B) \wedge M_x(B, A) = \infty .$$

The proof of this corollary will be given in section 5.2 .

The basic tool for finding conditions on $\Phi$ which are sufficient for (5.1) to hold is the equivalence of statements (5.9) (I) and (V) . This equivalence reduces the problem to that of investigating whether for some typical $\omega$ the ratios (5.5) have a finite upper or positive lower bound. Using two different techniques of estimating these ratios we will obtain two different sufficient conditions. But first let us discuss the consequences of Proposition (5.9) when (5.1) is satisfied. In particular, this will clarify the rôle played by the function $z(\cdot,\cdot,\cdot)$ in identifying the activity distribution in the extreme decomposition of canonical Gibbs measures.

For each $\emptyset \neq B \subset F$ we choose some $b_B \in B$ . Then for $\omega \in \Omega$ we let

$$B(\omega) = \{ b \in F : N(b, \omega) = \infty \}$$

and define

$$(5.13) \qquad z(a, \omega) = z(a, b_{B(\omega)}, \omega) / \sum_{c \in F} z(c, b_{B(\omega)}, \omega) \qquad (a \in F)$$

and

$$z(\omega) = z(\,.\,, \omega) \quad .$$

Since $z(b, b, .) = 1$, $z(.)$ is well-defined as soon as we agree upon any convention to define $\infty / \infty$. Clearly, the function $z(.)$ is tail-measurable.

(5.14) Theorem: *Suppose*

$$\text{ex } C \subset \bigcup_{z \in A} G(z) \quad .$$

*Then*

$$\text{ex } C = \bigcup_{z \in A} \text{ex } G(z) \quad .$$

*Moreover, this implies:*

(a) *For all* $\mu \in C$

$$E_\infty = F_\infty \qquad \mu\text{ - a. s.}$$

(b) *For all* $\mu \in C$ *and* $\mu$ *- a. a.* $\omega$ *we have* $z(\omega) \in A$ *and*

$$z(a, \omega) / z(b, \omega) = \lim_{\Lambda \uparrow S} Z_{\Lambda, N(\omega_\Lambda) + 1_b - 1_a}(\omega) / Z_{\Lambda, N(\omega_\Lambda)}(\omega)$$

*whenever* $a, b \in F$ *and* $N(b, \omega) = \infty$ .

(c) *Assume* $\mu \in C$ , $\Lambda \in S$ , *and* $\zeta \in \Omega_\Lambda$ . *Then* $\gamma_\Lambda^{z(.)}(\zeta | .)$ *is a version of the conditional probability* $\mu_\Lambda(\zeta | .)$ *with respect to* $F_{S \setminus \Lambda}$ .

(d) *For each* $z \in A$ *we have*

$$G(z) = \{ \mu \in C \; : \; z(.) \sim z \quad \mu \text{ - a. s.} \} \quad .$$

(e) *Let* $\mu \in C$ , $m$ *the distribution of* $z(.)$ *with respect to* $\mu$ , *and* $(z, A) \to \mu_z(A)$ *a regular version of the conditional probability* $\mu(A \mid z(.) = z)$ $(z \in A, A \in F)$ .

*Then*

$$m(\, z \in A \;:\; \mu_z \in G(z)\,) = 1$$

*and*

$$\mu = \int \mu_z \; m(dz) \quad .$$

*Moreover, if* $m_z$ *denotes the probability measure on* ex $G(z)$ *with barycentre* $\mu_z$ *then* $\int m(dz).m_z$ *is the probability measure on* ex $C$ *with barycentre* $\mu$ .

The proof of this Theorem is given in section 5.2. Now we introduce two conditions on $\Phi$ both of which will be seen to be sufficient for (5.1) .

(A) *There is a constant* $C < \infty$ *such that for infinitely many* $x \in S$

$$\max_{a,b \in F} \| E_x(a| \,.\,) - E_x(b| \,.\,) \| \leq C \quad .$$

Imposing stronger conditions on the interaction part of $\Phi$ we can weaken the requirement on the self-potential. We put

$$D(\Phi) = \max_{a,b \in F} \; \sup_{x \in S} \; \sum_{A \ni x} (|A| - 1) \max_{\omega \in \Omega_{A \setminus x}} | \Phi(A, a\omega) - \Phi(A, b\omega) | \quad .$$

(B) $D(\Phi)$ *is finite. Moreover, for any* $a, b \in F$ *the function* $x \to \Phi(x, a) - \Phi(x, b)$ *on* $S$ *is either bounded from above or from below. Finally*

$$\sum_{x \in S} \min_{a,b \in F} \exp [\, \Phi(x, a) - \Phi(x, b)\, ] = \infty \quad .$$

Notice that because of (5.12) the final condition in (B) is necessary for (5.1) when $D(\Phi) < \infty$ . Obviously, conditions (A) and (B) both imply (5.3) .

(5.15) <u>Theorem:</u> *Suppose* $\Phi$ *satisfies one of the conditions* (A) *and* (B) . *Then*

$$\text{ex } C \quad = \quad \bigcup_{z \in A} \quad \text{ex } G(z) \quad ,$$

*and the statements* (a) - (e) *of* (5.14) *hold.*

This will be proved in section 5.3 .

## 5.2 Conditional probabilities and the activity function

Let $\mu \in \text{ex } C$ be fixed. $\mu$ is supported on the set $\Omega_\mu$ introduced in (1.30) and also on

$$\Omega_\mu^+ \quad = \quad \{ \ \omega \in \Omega \ : \ \mu(X_\Lambda = \omega_\Lambda) > 0 \ \text{ for all } \ \Lambda \in S \ \} \quad .$$

For $x \in S$ , $a, b \in F$ , and $\omega \in \Omega$ we use the abbreviation

$$(5.16) \qquad D_x(\ a,\ b,\ \omega\ ) \quad = \quad \exp [\ -\ E_x(a|\omega) \ + \ E_x(b|\omega)\ ] \quad .$$

(5.17) Lemma: *Let* $\omega \in \Omega_\mu \cap \Omega_\mu^+$ , $a \in F$ , *and* $x \in V \in S$ . *For each* $W \in S$ *with* $W \supset V$ *we define*

$$q_W \quad = \quad D_x(\ \omega_x,\ a,\ \omega\ )\ \mu(\ X_W = a\omega_{W\setminus x}\ )\ /\ \mu(\ X_W = \omega_W\ ) \quad .$$

*Then there are numbers* $\delta(W) \geq 0$ *such that* $\lim_{W \uparrow S} \delta(W) = 0$ *and*

$$e^{-\delta(W)} \ q_W \ \leq \ \limsup_{\Lambda \uparrow S} \ z_{\Lambda,W,V}\ (\ a,\ \omega_x,\ \omega\ ) \ \leq \ e^{\delta(W)} \ q_W$$

*whenever* $V \subset W \in S$ .

(Recall that the partition function ratio $z_{\Lambda,W,V}(\cdot\ ,\cdot\ ,\cdot)$ was defined in (5.5) . )

Proof: By the assumption on $\omega$ we have

$$q_W = \lim_{\Lambda \uparrow S} q_{\Lambda,W}$$

where

$$q_{\Lambda,W} = D_X( \omega_X, a, \omega ) \; \gamma_{\Lambda,L}( X_W = a\omega_{W \setminus X} | \omega ) \; / \; \gamma_{\Lambda,L}( X_W = \omega_W | \omega )$$

whenever $W \subset \Lambda \in S$. Here we have written $L$ for $N(\omega_\Lambda)$. Now we observe that

$$D_X( \omega_X, a, \omega ) = \exp [ E_V( a\omega_{V \setminus X} | \omega ) - E_V( \omega_V | \omega ) ] .$$

Consequently, $q_{\Lambda,W}$ is the ratio of

$$(5.18) \qquad \sum_{\zeta \in \Omega_{\Lambda \setminus W, L - N(a\omega_{W \setminus X})}} \exp [ E_V( a\omega_{V \setminus X} | \omega ) - E_\Lambda( a\omega_{W \setminus X} \zeta | \omega ) ]$$

and the same expression with a replaced by $\omega_X$. Similarly, $z_{\Lambda,W,V} ( a, \omega_X, \omega )$ is the ratio of the partition function

$$(5.19) \qquad \sum_{\zeta \in \Omega_{\Lambda \setminus W, L - N(a\omega_{W \setminus X})}} \exp [ - E_{\Lambda \setminus W}( \zeta | \alpha_V \, \omega_{S \setminus V} ) ]$$

and the corresponding term with a replaced by $\omega_X$. (Notice that $L - N( a\omega_{W \setminus X} ) = N(\omega_{\Lambda \setminus W}) + 1_{\omega_X} - 1_a$ ). We show that the exponentials in (5.18) and (5.19) differ only by a term $C$ which is independent of $a$ and $\zeta$ and by two terms which are bounded by

$$\Delta(W) = \sup \{ \, | E_V(\zeta|\eta) - E_V(\zeta|\eta') | \; : \; \zeta \in \Omega_V \, , \; \eta, \eta' \in \Omega \, , \; \eta W = \eta' W \} .$$

By the definition of a potential, $E_V(\zeta | \cdot )$ is uniformly continuous on $\Omega$. Thus

$\Delta(W) \downarrow 0$  as  $W \uparrow S$ .  For the constant term we choose

$$C = E_V( \alpha_V|\omega_W \alpha_{\Lambda \smallsetminus W} \omega_{S \smallsetminus \Lambda} ) - E_\Lambda( \alpha_V \omega_{W \smallsetminus V} \alpha_{\Lambda \smallsetminus W}|\omega ) + E_{\Lambda \smallsetminus W}( \alpha_{\Lambda \smallsetminus W}|\alpha_V \omega_{S \smallsetminus V} ) .$$

Using the decomposition

$$\sum_{A \cap \Lambda \neq \emptyset} = \sum_{A \cap V \neq \emptyset} + \sum_{A \subset W \smallsetminus V} + \sum_{\substack{A \cap (\Lambda \smallsetminus W) \neq \emptyset \\ A \cap V = \emptyset}}$$

we then obtain the estimate

$$| E_V( a\omega_{V \smallsetminus x}|\omega ) - E_\Lambda( a\omega_{W \smallsetminus x} \zeta|\omega ) + E_{\Lambda \smallsetminus W}( \zeta|\alpha_V \omega_{S \smallsetminus V} ) - C |$$

$$= | E_V( a\omega_{V \smallsetminus x}|\omega ) - E_V( a\omega_{V \smallsetminus x}|\omega_W \zeta\omega_{S \smallsetminus \Lambda} )$$

$$+ \sum_{\substack{A \cap (\Lambda \smallsetminus W) \neq \emptyset \\ A \cap V \neq \emptyset}} [ \Phi( A, \alpha_V \omega_{W \smallsetminus V} \zeta\omega_{S \smallsetminus \Lambda} ) - \Phi( A, \alpha_V \omega_{W \smallsetminus V} \alpha_{\Lambda \smallsetminus W} \omega_{S \smallsetminus \Lambda} ) ] |$$

$$\leq \Delta(W) + | E_V( \alpha_V|\omega_W \zeta\omega_{S \smallsetminus \Lambda} ) - E_V( \alpha_V|\omega_W \alpha_{\Lambda \smallsetminus W} \omega_{S \smallsetminus \Lambda} ) |$$

$$\leq 2 \Delta (W) .$$

This gives us the result

$$e^{-4\Delta(W)} q_{\Lambda,W} \leq Z_{\Lambda,W,V} ( a, \omega_x, \omega ) \leq e^{4\Delta(W)} q_{\Lambda,W} .$$

Taking the limit  $\Lambda \uparrow S$  we complete the proof.  ⌟

Let us write  $z(a, b)$  instead of  $z( a, b, \mu )$  $( a, b \in F )$ .

(5.20) **Proposition:** *Let*  $x \in S$ ,  $a \in F$ .  *Then for*  $\mu$ - a. a. $\omega$  *the equation*

$$\mu_x(a|\omega) = z(a, \omega_x)\ D_x(a, \omega_x, \omega)\ \mu_x(\omega_x|\omega)$$

*holds.*

Proof: Let $\omega \in \Omega_\mu^+$ and $q_W$ be as in (5.17). Then we can write

$$q_W = D_x(\omega_x, a, \omega)\ \mu(X_x = a|F_{W\setminus x})(\omega) / \mu(X_x = \omega_x|F_{W\setminus x})(\omega) .$$

Taking the limit $W \uparrow S$ we can use the martingale convergence theorem and Lemma (5.17). This gives

$$D_x(\omega_x, a, \omega)\ \mu_x(a|\omega) = z_V(a, \omega_x, \omega)\ \mu_x(\omega_x|\omega)$$

for $\mu$ - a. a. $\omega$ and all $V \ni x$. Here

$$z_V(\cdot, \cdot, \cdot) = \limsup_{W\uparrow S}\ \limsup_{\Lambda\uparrow S}\ z_{\Lambda,W,V}(\cdot, \cdot, \cdot) .$$

The desired conclusion follows by taking the trivial limit $V \uparrow S$. ┘

Now we can start with the proof of proposition (5.9). The implication (I) ⇒ (II) is trivial since $\gamma_x^z(a|\cdot) > 0$ whenever $z(a) > 0$. In order to deduce (III) from (II) we observe that (III) is equivalent to

$$\sup_{x\in S} \int d\mu\ 1_{\{X_x=a\}}\ \mu_x(b|\cdot) > 0 \qquad (a, b \in F_\mu)$$

and therefore to (5.10). But obviously (II) implies (5.10). The equivalence of the statements (III), (IV), and (V) follows from

(5.21) Lemma: *Suppose* $a, b \in F_\mu$. *Then the inequality*

$$\mu(X_x = a ,\ \mu_x(b|\cdot) > 0) > 0 .$$

*holds for all* $x \in S$ *provided it holds for some* $x \in S$ . *This happens iff* $z(b, a) > 0$ . *In particular,* $z(b, a) > 0$ *iff* $z(a, b) > 0$ .

<u>Proof:</u> If $z(b, a) = 0$ then (5.20) shows that for all $x$ $\mu_x(b| \, .\,) = 0$ $\mu$ - a. s. on $\{X_x = a\}$ . Conversely, assume $z(b, a) > 0$ . Then again (5.20) implies that for all $x \in S$ and $\mu$ - a. a. $\omega$ $\mu_x(b|\omega) > 0$ whenever $\mu_x(a|\omega) > 0$. But clearly $\mu_x(a| \, .\,) > 0$ $\mu$ - a. s. on $\{X_x = a\}$ . Thus

$$\mu( X_x = a \; , \; \mu_x(b| \, .\,) > 0 ) \quad = \quad \mu(X_x = a) \; > \; 0 \quad .$$

The final assertion comes from the observation that the inequality just proved is equivalent to the inequality

$$\int d\mu \quad \mu_x(a| \, .\,) \;\; \mu_x(b| \, .\,) \;\; > \;\; 0$$

which is symmetric in $a$ and $b$ . ┘

The final implication (IV) ⇒ (I) is the content of

(5.22) <u>Lemma:</u> *Suppose* $z(a, b) > 0$ *for all* $a$ , $b \in F_\mu$ . *Then the following statements hold:*

(i) $z(a, b) < \infty$ *whenever* $a \in F$ , $b \in F_\mu$ .

(ii) *For each* $a \in F$ , $b, c \in F_\mu$ *we have*

$z(a, b) \;\; z(b, c) \;\; = \;\; z(a, c)$ .

*In particular,* $z( \, . \, , b) \sim z( \, . \, , c)$ .

(iii) $\mu \in \mathcal{G}( z( \cdot \, , b) )$ *for any* $b \in F_\mu$ .

<u>Proof:</u> 1. Let $a \in F$ , $b \in F_\mu$ , $x \in S$ . Then $\mu(X_x = b) > 0$ , and $\mu$ - a. s. on $\{X_x = b\}$ we have $\mu_x(b| \, .\,) > 0$ and, because of (5.20) ,

$$z(a, b) \quad = \quad \mu_x(a| \, .\,) \;\; D_x(b, a, .\,) \; / \; \mu_x(b| \, .\,) \;\; < \;\; \infty \quad .$$

2. Again let $a \in F$, $b \in F_\mu$, $x \in S$. We show that for all $c \in F_\mu$

$$\mu(\{X_x = c\} \cap A) > 0$$

where

$$A = \{\mu_x(a|.) = z(a, b)\ D_x(a, b, .)\ \mu_x(b|.)\} .$$

Clearly $A \in F_{S \setminus x}$, and (5.20) gives

$$\{X_x = b\} \setminus A = \emptyset \qquad \mu\text{ - a. s.}$$

Therefore if

$$\{X_x = c\} \cap A = \emptyset \qquad \mu\text{ - a. s.}$$

then

$$\{X_x = b\} = \{X_x = b \text{ or } c\} \cap A \qquad \mu\text{ - a. s.}$$

and thus

$$\mu_x(b|.) = 1_A [\mu_x(b|.) + \mu_x(c|.)] \qquad \mu\text{ - a. s.}$$

This leads to the conclusion

$$0 = 1_{\{X_x=b\}}\ 1_{\Omega \setminus A}\ \mu_x(b|.) =$$

$$= 1_{\{X_x = b\}}\ 1_A\ \mu_x(c|.) = 1_{\{X_x = b\}}\ \mu_x(c|.) \qquad \mu\text{ - a. s.}$$

Combining this with Lemma (5.21) we see that $z(c, b) = 0$ in contradiction to our assumption.

3. Next we prove $z(a, b)\ z(b, c) = z(a, c)$ whenever $a \in F$, $b, c \in F_\mu$. Indeed, we see from 2. and Proposition (2.20) that for any $x$ the following

relations hold simultaneously with positive probability:

(5.23 a) $\mu_x(a|\,.\,) = z(a, b)\ D_x(a, b, .)\ \mu_x(b|\,.\,)$

(5.23 b) $\mu_x(b|\,.\,) = z(b, c)\ D_x(b, c, .)\ \mu_x(c|\,.\,)$

(5.23 c) $\mu_x(a|\,.\,) = z(a, c)\ D_x(a, c, .)\ \mu_x(c|\,.\,)$

(5.23 d) $\mu_x(c|\,.\,) > 0$ .

Solving for $z$ the assertion follows.

4. Now let $a \in F$, $b \in F_\mu$, $x \in S$. We show that (5.23 a) holds with probability 1. Because of $\mu(X_x \in F_\mu) = 1$ we need only show that (5.23 a) holds a. s. on $\{X_x = c\}$ for each $c \in F_\mu$. However, (5.20) asserts that for all $c \in F_\mu$ (5.23 b & c) are a. s. satisfied on $\{X_x = c\}$. Combining these two identities we get

$$\mu_x(a|\,.\,) = z(a, c)\ z(b, c)^{-1}\ D_x(a, b, .)\ \mu_x(b|\,.\,)$$

a. s. on $\{X_x = c\}$. The claim now follows from step 3. which gives

$$z(a, c) / z(b, c) = z(a, c)\ z(c, b) = z(a, b) .$$

5. Suppose $b \in F_\mu$. Then $\mu \in G(\ z(\,.\,, b)\ )$. Indeed, for all $x \in S$ and $a \in F$ we obtain from 4. that $\mu_x(b|\,.\,) > 0$ a. s. and

$$\begin{aligned}\mu_x(a|\,.\,) &= \mu_x(a|\,.\,) / \sum_{a'\in F} \mu_x(a'|\,.\,) \\ &= z(a, b)\ D_x(a, b, .)\ \mu_x(b|\,.\,) / \sum_{a'\in F} z(a', b)\ D_x(a', b, .)\ \mu_x(b|.) \\ &= \gamma_x^z(a|\,.\,) \quad \text{a. s.}\end{aligned}$$

where $z = z(\,.\,, b)$. Dobrushin (1968 a) has pointed out that this already implies $\mu \in G(z)$. We reproduce his argument. We have to show that for all $\Lambda \in S$

and $\zeta \in \Omega_\Lambda$

$$\mu_\Lambda(\zeta | \,.\,) = \gamma_\Lambda^z(\zeta | \,.\,) \quad \text{a. s.} \tag{5.24}$$

By hypothesis, this is true if $|\Lambda| = 1$ . In particular, we see that

$$\begin{aligned} F_\mu &= \{ a \in F : \mu( \mu_x(a | \,.\,) > 0 ) > 0 \text{ for some } x \} \\ &= \{ a \in F : z(a) > 0 \} . \end{aligned}$$

Moreover, for all $\Lambda \in S$ and $\zeta \in F_\mu^\Lambda$ we have $\mu(X_\Lambda = \zeta) > 0$ . (Clearly this is true if $|\Lambda| = 1$ , and for general $\Lambda$ it follows from an induction argument based on the equation

$$\mu(X_x = \zeta) = \int_{\{X_{\Lambda \setminus x} = \zeta_{\Lambda \setminus x}\}} \mu_x(\zeta_x | \,.\,) \, d\mu$$

where $x \in \Lambda$ . ) Now we prove (5.24) by induction on $|\Lambda|$ . Assuming $|\Lambda| > 1$ we can write $\Lambda = V \cup W$ with $V \cap W = \emptyset$ , $V \neq \emptyset$ , $W \neq \emptyset$ . Then the consistency (1.17) of grand canonical Gibbs distributions gives for arbitrary $\zeta,\ \alpha \in F_\mu^\Lambda$ and $\omega \in \Omega$

$$\begin{aligned} &\gamma_\Lambda^z(\alpha|\omega) \,/\, \gamma_\Lambda^z(\zeta|\omega) = \\ &= [\, \gamma_\Lambda^z(\alpha|\omega) \,/\, \gamma_\Lambda(\zeta_V\, \alpha_W|\omega) \,]\, [\, \gamma_\Lambda(\zeta_V\, \alpha_W|\omega) \,/\, \gamma_\Lambda^z(\zeta|\omega) \,] \\ &= [\, \gamma_V^z(\alpha_V|\alpha_W\, \omega_{S\setminus W}) \,/\, \gamma_V^z(\zeta_V|\alpha_W\, \omega_{S\setminus W}) \,]\, [\, \gamma_W^z(\alpha_W|\zeta_V\, \omega_{S\setminus V}) \,/\, \gamma_W^z(\zeta_W|\zeta_V\, \omega_{S\setminus V}) \,] \\ &= (\, \gamma_1 \,/\, \gamma_2 \,)\, (\, \gamma_3 \,/\, \gamma_4 \,) . \end{aligned}$$

Now the induction hypothesis asserts that

$$\gamma_1 = \mu_V(\, \alpha_V|\alpha_W\, \omega_{S\setminus W} \,) \tag{5.25}$$

$$= \lim_{\Delta \uparrow S \setminus \Lambda} \mu( X_V = \alpha_V \mid X_W = \alpha_W , \; X_\Delta = \omega_\Delta )$$

for $\mu$ - a. a. $\omega \in \{X_W = \alpha_W\}$ . Moreover,

$$\mu_W(\alpha_W | \,.\,) = \gamma_W^z(\alpha_W | \,.\,) > 0 \qquad \mu \text{ - a. s.}$$

and thus (5.25) holds for $\mu$ - a. a. $\omega \in \Omega$ . The factors $\gamma_2$ , $\gamma_3$ , $\gamma_4$ can be treated similarly. Therefore we have for $\mu$ - a. a. $\omega$

$$( \gamma_1 / \gamma_2 ) ( \gamma_3 / \gamma_4 ) = \lim_{\Delta \uparrow S \setminus \Lambda} \mu( X_\Lambda = \alpha , \; X_\Delta = \omega_\Delta ) / \mu( X_\Lambda = \zeta , X_\Delta = \omega_\Delta ).$$

Summing up over $\alpha$ we finally obtain

$$\gamma_\Lambda^z(\zeta | \omega) = \lim_{\Delta \uparrow S \setminus \Lambda} \mu( X_\Lambda = \zeta \mid X_\Delta = \omega_\Delta ) = \mu_\Lambda(\zeta | \omega) .$$

Since this equation is trivially satisfied whenever $\zeta \in \Omega_\Lambda \setminus F_\mu^\Lambda$ , the proof of (5.24) is complete. ┘

Let us finish the proof of Proposition (5.9) by verifying the final assertions. The first was already shown in (5.11) . Next, if $\mu \in \mathcal{G}(z)$ then (5.20) gives $z(a, b, \mu) = z(a)/z(b)$ $(a \in F, b \in F_\mu)$ . Finally, the last statements of (5.9) are proved in (5.22) .

Now we give the

(5.26) <u>Proof of Corollary (5.12):</u>

Suppose there are disjoint sets $A, B \subset F$ with

$$\sum_{x \in S} M_x(A, B) \wedge M_x(B, A) < \infty .$$

We put $z = 1_{A \cup B}$ and show

$$\text{ex } C \cap G(z) = \emptyset .$$

Let $x \in S$ . Then

$$\gamma_x^z(A| \,.\,) = [\, 1 + (\sum_{a\in A} \exp [\, - E_x(a| \,.\,) \,] \,/ \sum_{b\in B} \exp [\, - E_x(b| \,.\,) \,] \,)^{-1} \,]^{-1}$$

and thus for each $\mu \in G(z)$

$$\mu( X_x \in A ) = \int_{(A\cup B)^S} d\mu \;\; \gamma_x^z(A| \,.\,)$$

$$\leq [\, 1 + M_x(A, B)^{-1} \,]^{-1} \leq M_x(A, B) .$$

Similarly,

$$\mu( X_x \in B ) \leq M_x(B, A) .$$

We define $S_1 = \{ x \in S : M_x(A, B) \leq M_x(B, A) \}$ and $S_2 = S \setminus S_1$ . Then our assumption gives

$$N(A, S_1) = \sum_{a\in A} N(a, X_{S_1}) < \infty \qquad \mu - \text{a. s.}$$

and

$$N(B, S_2) < \infty \qquad \mu - \text{a. s.}$$

Thus

$$D = 1_{\{N(F\setminus(A\cup B),S)=0\}} \;\; [\, N(A, S_1) - N(B, S_2) \,]$$

is $\mu$ - a. s. well-defined and finite. Clearly, $D$ is invariant when finitely many coordinates are permuted. Hence $D$ is measurable with respect to $E_\infty$ . Therefore for each $\mu \in \text{ex } C \cap G(z)$ there is an integer $k$ such that $\mu(D = k) = 1$ .

Consequently, for any $x \in S$, any $x \in S_1$, we have

$$\int d\mu \; 1_{\{X_x \in A\}} \; \mu_x(B|\bullet) =$$

$$= \int d\mu \; 1_{\{X_x \in A\}} \; 1_{\{N(A,S_1 \setminus x) - N(B,S_2) = k-1\}} \; \mu_x(B|\bullet)$$

$$\leq \int d\mu \; 1_{\{N(A,S_1 \setminus x) - N(B,S_2) = k-1\}} \; 1_{\{X_x \in B\}} = 0 .$$

Thus, (5.9) (III) is violated. This contradicts the fact that (5.9) (I) holds. ⌋

(5.27) Proof of Theorem (5.14):

1. For each $\mu \in C$ we have $\mu( z(.) \in A ) = 1$. Clearly, it is sufficient to verify this for $\mu \in \text{ex } C$. Then by assumption (5.9) (I) holds, and we obtain from (5.9)

$$B\,(.) = F_\mu \qquad \mu\text{ - a. s.}$$

and $z( a, b_{F_\mu}, \mu ) < \infty$ for all $a \in F$. This means that for $\mu$ - a. a. $\omega$ and all $a \in F$ $z(a, \omega)$ (as defined in (5.13) ) is finite. Thus $z(.) \in A$ $\mu$ - a. s.

2. If $\mu \in \text{ex } C$ then $\mu \in G( z(\mu) )$. Indeed, our assumption guarantees that all statements on (5.9) are satisfied by $\mu$. Thus $\mu \in G( z( . , b_{F_\mu}, \mu) )$. But the argument in step 1. above shows that $z(\mu) \sim z( . , b_{F_\mu}, \mu)$. Hence $\mu \in G( z(\mu) )$.

3. Now we prove that for all $\mu \in C$, $a, b \in F$ and $\mu$ - a. a. $\omega$ $z(a, \omega) / z(b, \omega)$ can be obtained as a limit of ratios of canonical partition functions. (Together with 1. this will complete the proof of (b). ) Obviously, we can assume $\mu \in \text{ex } C$, i. e., $\mu \in G(z)$ where $z = z(\mu)$ is a probability vector on $F$. We add an extra site $0$ to $S$ and put $\underline{S} = S \cup \{0\}$. All quantities which refer to this enlarged set will be underlined. We define a potential $\underline{\Phi}$ on $\underline{S}$ by

$$\underline{\Phi}(\ \underline{A},\ \underline{\omega}\ ) \ = \begin{cases} \Phi(\ \underline{A},\ \underline{\omega}_S\ ) & \text{if} \quad \underline{A} \subset S \\ 0 & \text{otherwise} \end{cases}$$

and denote by $\underline{G}(z)$ and $\underline{C}$ the corresponding sets of grand canonical or canonical Gibbs measures on $\underline{\Omega} = F \times \Omega$ . We put

$$\underline{\mu} \ = \ z \otimes \mu \quad .$$

It is easy to check that $\underline{\mu} \in \underline{G}(z) \subset \underline{C}$ . Moreover, $\underline{\mu} \in \text{ex}\ \underline{C}$ . In order to see this we choose a set $\underline{A} \in \underline{E}_\infty$ and prove $\underline{\mu}(\underline{A}) = 0$ or $1$ . Suppose $\underline{\mu}(\underline{A}) > 0$ . Then for each $a$ with $z(a) > 0$ we have

$$\underline{\mu}(\ \{X_0 = a\} \cap \underline{A}\ ) \ > \ 0 \quad .$$

Indeed, $\{\ N(a,\bullet\,) > 0\ \} \in E_\infty$ and $\mu$ is trivial on $E_\infty$ . Thus

$$\underline{\mu}(\ N(a,\ X_S) > 0\ ) \ = \ \mu(\ N(a,.\,) > 0\ ) \ = \ 1$$

and therefore

$$\underline{\mu}(\ \bigcup_{x \in S}\ \{X_x = a\} \cap \underline{A}\ ) \ = \ \underline{\mu}(A) \ > \ 0 \quad .$$

Consequently, there is an $x \in S$ such that $\underline{\mu}(\ \{X_x = a\} \cap \underline{A}\ ) > 0$ . Moreover, the conditional probability $\underline{\nu} = \underline{\mu}(\bullet \mid \underline{A}\ )$ is well-defined, and $\underline{A} \in \underline{E}_\infty$ implies $\underline{\nu} \in \underline{C}$ . We have just shown $a \in F_{\underline{\nu}}$ . Hence (5.8) implies $\underline{\mu}(\ X_0 = a \mid \underline{A}\ ) > 0$ , as desired. Now let $A_a = \{\ \omega \in \Omega\ :\ (a,\omega) \in \underline{A}\ \}$ . Clearly, $A_a \in E_\infty$ . Furthermore,

$$z(a)\ \ \mu(A_a) \ = \ \underline{\mu}(\ \{X_0 = a\} \cap \underline{A}\ ) \ > \ 0$$

and thus $\mu(A_a) > 0$ . Since $\mu$ is extreme, $\mu(A_a) = 1$ . Therefore we obtain

$$\underline{\mu}\,(\underline{A}) = \sum_{a:z(a)>0} z(a)\ \mu(A_a) = \sum_{a\in F} z(a) = 1 .$$

This completes the proof that $\underline{\mu} \in \text{ex}\ \underline{C}$ . In order to show the desired limit representation of $z(\omega)$ we apply Theorem (1.32) as follows. Let $a \in F$ , $b \in F_\mu = F_{\underline{\mu}}$ . Then for $\underline{\mu}$ - a. a. $\underline{\omega} = (b, \omega) \in \{X_0 = b\}$ we get (writing $\underline{\Lambda} = \Lambda \cup \{0\}$)

$$z(\,a,\,b,\,\mu\,) = z(a)\,/\,z(b) = \underline{\mu}(\,X_0 = a\,)\,/\,\underline{\mu}(\,X_0 = b\,)$$

$$= \lim_{\Lambda\uparrow S} \underline{\gamma}_{\underline{\Lambda},N(\omega_\Lambda)+1_b}\,(\,X_0 = a \mid \underline{\omega}\,)\,/\,\underline{\gamma}_{\underline{\Lambda},N(\omega_\Lambda)+1_b}\,(\,X_0 = b \mid \underline{\omega}\,)$$

$$= \lim_{\Lambda\uparrow S} Z_{\Lambda,N(\omega_\Lambda)+1_b-1_a}(\omega)\,/\,Z_{\Lambda,N(\omega_\Lambda)}(\omega) .$$

Thus (5.14) (b) is completely proved.

4. The next step is to verify (5.14) (c) . We let $C'$ denote the set of all $\mu \in C$ which satisfy

$$\mu_\Lambda\,(\zeta\,|\,.\,) = \gamma_\Lambda^{z(\cdot)}\,(\zeta\,|\,.\,) \qquad \mu\text{ - a. s.}$$

whenever $\Lambda \in S$ and $\zeta \in \Omega_\Lambda$ . Clearly, $C'$ is convex. Thus in order to show $C' = C$ it is sufficient to prove $\text{ex}\ C \subset C'$ . This is a simple consequence of step 2.

5. Now let us prove (5.14) (a) . Let $\mu \in C$ , $A \in E_\infty$ . We can assume $\mu(A) > 0$ . (Otherwise $A = \emptyset \in F_\infty$ $\mu$ - a. s.) . We put $f = 1_A\,/\,\mu(A)$ , and for any $\Lambda \in S$ we let $f_\Lambda$ denote the conditional expectation of $f$ with respect to $\mu$ and $F_{S\setminus\Lambda}$ . We need only show that $f = f_\Lambda$ $\mu$ - a. s. for all $\Lambda \in S$ . Indeed, this implies

$$A = \{ \mu(A|F_\infty) = 1 \} \in F_\infty \qquad \mu - a.\ s.$$

Therefore we fix a set $\Lambda \in S$ and choose $\zeta \in \Omega_\Lambda$ , $B \in F_{S \setminus \Lambda}$ . Then from (5.14) (c) (applied to $f\mu = \mu(\cdot \mid A) \in C$ and $\mu$ ) we obtain

$$(f\mu)(\{X_\Lambda = \zeta\} \cap B) = \int_B \gamma_\Lambda^{z(\cdot)}(\zeta \mid .)\, f\, d\mu =$$

$$= \int_B \gamma_\Lambda^{z(\cdot)}(\zeta \mid .)\, f_\Lambda\, d\mu = \int_B 1_{\{X_\Lambda = \zeta\}}\, f_\Lambda\, d\mu$$

$$= (f_\Lambda \mu)(\{X_\Lambda = \zeta\} \cap B) \quad .$$

This gives $f\mu = f_\Lambda \mu$ .

6. If $\mu \in \text{ex}\, G(z)$ for some $z \in A$ then $\mu \in \text{ex}\, C$ . This is an immediate consequence of (5.14) (a) and the characterizations of ex $G(z)$ and ex $C$ by zero-one laws for $F_\infty$ and $E_\infty$ respectively.

7. In order to prove (5.14) (d) we observe that (because of (5.14) (c)) $\mu \in G(z)$ whenever $\mu \in C$ and $z(.) \sim z$ $\mu$ - a. s. Conversely, $\mu(z(.) \sim z) = 1$ for any $\mu \in G(z)$ . For if $\mu \in \text{ex}\, G(z)$ then $\mu \in \text{ex}\, C$ , and (5.9) gives for all $a \in F$ , $b \in F_\mu$

$$z(a) / z(b) = z(a, b, \mu) = z(a, b_{F_\mu}, \mu) / z(b, b_{F_\mu}, \mu)$$

$$= z(a, \mu) / z(b, \mu) \quad .$$

Thus $z \sim z(\mu)$ .

8. Finally we prove (5.14) (e) . The martingale convergence theorem asserts that for m - a. a. $z \in A$ $\mu_z$ is the weak limit of conditional probabilities of the form $\mu(\cdot \mid z(.) \in I)$ , where $I \subset A$ and $m(I) > 0$ . Since $z(.)$ is tail-

measurable, $\mu( . \mid z(.) \in I ) \in C$ . Thus the compactness of $C$ gives

$$m( z \in A : \mu_z \in C. ) = 1 .$$

Clearly,

$$m( z \in A : \mu_z( z(.) = z) = 1 ) = 1 .$$

Therefore (5.14) (d) implies

$$m( z \in A : \mu_z \in G(z) ) = 1 .$$

The remaining assertions are obvious. ⌟

## 5.3 Estimating the activity function

In this section we will use the conditions (A) and (B) in order to estimate the ratios (5.5) of canonical partition functions. These estimates are based on the following

(5.28) Remark: *Suppose* $\Lambda \in S$ , $\omega \in \Omega$ , $a, b \in F$ *and* $L \in A_\Lambda$ . *Then*

$$( L(b) + 1 ) \; Z_{\Lambda,L + 1_b - 1_a}(\omega) \,/\, Z_{\Lambda,L}(\omega) =$$

$$= \sum_{\zeta\in\Omega_{\Lambda,L}} \gamma_{\Lambda,L}(\zeta|\omega) \sum_{x\in\Lambda:\zeta_x=a} D_x( b, a, \zeta\,\omega_{S\setminus\Lambda} ) .$$

Proof: Each configuration in $\Omega_{\Lambda,L + 1_b - 1_a}$ can be obtained in exactly $L(b) + 1$ different ways by changing the value of a configuration $\zeta \in \Omega_{\Lambda,L}$ at a site $x$ with $\zeta_x = a$ into $b$ . $D_x( b, a, \zeta\omega_{S\setminus\Lambda} )$ is the energy change caused by this replacement. ⌟

Now let us assume that condition (A) is satisfied. This condition can be expressed as follows: If $m_x$ is defined by (5.2) then there is a constant $c > 0$ such that

$$S_c = \{ x \in S : m_x \geq c \}$$

is infinite. This condition implies that we can construct a sequence $\Lambda \uparrow S$ such that

(5.29) $$|\Lambda \cap S_c| / |\Lambda| \to 1 .$$

We can (and will) assume that in the definition (5.6) of $z(a, b, \omega)$ and in (5.30) below $\Lambda$ runs through this particular sequence.

We consider a fixed measure $\mu \in \text{ex } C$.

(5.30) Lemma: *If the potential $\Phi$ satisfies condition (A) then for all $a \in F_\mu$ and $\omega \in \Omega_\mu$ we have*

$$\rho_-(a, \omega) = \liminf_{\Lambda \uparrow S} N(a, \omega_\Lambda) / |\Lambda| > 0 .$$

Proof: Let us assume $\rho_-(a, \omega) = 0$ for some $a \in F_\mu$ and $\omega \in \Omega_\mu$. Let $x \in S_c$. For given $\Lambda \in \mathcal{S}$ we introduce the short-hand notations $L = N(\omega_\Lambda)$, $L_a = L - 1_a$, and $\Lambda_x = \Lambda \setminus x$. Then

$$\frac{1 - \mu(X_x = a)}{\mu(X_x = a)} = \lim_{\Lambda \uparrow S} \sum_{b \neq a} \frac{\gamma_{\Lambda,L}(X_x = b|\omega)}{\gamma_{\Lambda,L}(X_x = a|\omega)} =$$

$$= \lim_{\Lambda \uparrow S} \sum_{b \neq a} \frac{\sum_{\zeta \in \Omega_{\Lambda_x, L_b}} D_x(b, a, \zeta\omega_{S \setminus \Lambda}) \exp[-E_{\Lambda_x}(\zeta|a\omega_{S \setminus x})]}{\sum_{\zeta \in \Omega_{\Lambda_x, L_a}} \exp[-E_{\Lambda_x}(\zeta|a\omega_{S \setminus x})]}$$

$$\geq \; m_x \limsup_{\Lambda \uparrow S} \sum_{b \neq a} Z_{\Lambda_x, L_b}(a\omega_{S \setminus x}) \,/\, Z_{\Lambda_x, L_a}(a\omega_{S \setminus x})$$

$$\geq \; c \limsup_{\Lambda \uparrow S} L(a)^{-1} \sum_{\zeta \in \Omega_{\Lambda_x, L_a}} \gamma_{\Lambda_x, L_a}(\zeta \mid a\omega_{S \setminus x}) \sum_{y : \zeta_y \neq a} m_y \quad .$$

The final step is a consequence of (5.28). Now we observe that for each $\zeta \in \Omega_{\Lambda_x, L_a}$

$$|\{ y \in \Lambda_x : \zeta_y = a \text{ or } y \notin S_c \}| \leq L_a(a) + |\Lambda \setminus S_c|$$

and therefore

$$\sum_{y \in \Lambda_x : \zeta_y \neq a} m_y \geq c( |\Lambda \cap S_c| - L(a) ) \quad .$$

Hence we see that $( 1 - \mu(X_x = a) ) / \mu(X_x = a)$ has the lower bound

$$c^2 \limsup_{\Lambda \uparrow S} ( |\Lambda \cap S_c| / L(a) - 1 ) = \infty$$

which is impossible since $a \in F_\mu$ . ┘

Notice that for each $a \in F$ $\rho_-(a, .)$ is tail-measurable and thus $\mu$ - a. s. constant.

(5.31) <u>Proposition:</u> *Suppose condition* (A) *is satisfied. Then for all* $a, b \in F_\mu$ *we have*

$$z( a, b, \mu ) \geq c \; \rho_-(a, \mu) / \rho_-(b, \mu) > 0 \quad .$$

<u>Proof:</u> We may assume $a \neq b$ . Let $\omega \in \Omega_\mu$ . Then (5.28) gives

$$(5.32) \qquad (L(b)+1)\; z_{\Lambda,W,V}\,(a,b,\omega)$$

$$\geq \sum_{\zeta\in\Omega_{\Lambda\setminus W,L}} \gamma_{\Lambda\setminus W,L}\,(\zeta\,|\,\alpha_V\,\omega_{S\setminus V}) \sum_{x:\zeta_x=a} m_x$$

whenever $V \subset W \subset \Lambda \in S$ and $L = N(\omega_{\Lambda\setminus W})$ . Moreover, for each $\zeta \in \Omega_{\Lambda\setminus W,L}$ we have

$$|\{ x \in \Lambda \setminus W \;:\; \zeta_x \neq a \text{ or } x \notin S_c \}| \leq |\Lambda| - N(a,\omega_\Lambda) + |\Lambda \setminus S_c|$$

Thus the r. h. s. of (5.32) has the lower bound

$$c\;(\,N(a,\omega_\Lambda) - |W| - |\Lambda \setminus S_c|\,) \quad .$$

Given $\varepsilon > 0$ this expression is greater than

$$c\;|\Lambda|\;(\,\rho_-(a,\omega) - \varepsilon\,)$$

provided $\Lambda$ is sufficiently large. Now the proposition is evident. ┘

Now we consider the case when condition (B) is satisfied.

(5.33) Lemma: *Suppose* $\sum_{x\in S} m_x = \infty$ . *Then*

$$\mu(\,N(a,.) = \infty\,) = 1$$

*for all* $a \in F_\mu$ .

Proof: Assume $\mu(\,N(a,.) < \infty\,) > 0$ for some $a \in F_\mu$ . Then there is an $\omega \in \Omega_\mu$ with $1 \leq N(a,\omega) < \infty$ , and an approximation argument similar to that in the proof of (5.30) leads to the contradiction $(1 - \mu(X_x = a))/\mu(X_x = a) = \infty$

whenever $m_x > 0$ . ┘

(5.34) Proposition: *If condition* (B) *is satisfied then we have*

$$z( a, b, \mu ) \geq 1 / z( b, a, \mu )$$

*for any* $a, b \in F_\mu$ .

Proof: We choose $\omega \in \Omega$ and $a, b \in F$ such that $N(a, \omega) = N(b, \omega) = \infty$ and $\Delta \in S$ so large that $N(a, \omega_\Delta)\, N(b, \omega_\Delta) \geq 1$ . We put $L = N(\omega_\Delta)$ and $K = L - 1_b + 1_a$ and fix an arbitrary boundary condition $\eta \in \Omega$ . (We think of the case when $\Delta = \Lambda \setminus W$ and $\eta = \alpha_V\, \omega_{S \setminus V}$ for three sets $V \subset W \subset \Lambda \in S$ .) Applying Jensen's inequality to (5.28) we obtain

$$[\, L(b)\; Z_{\Delta,L}\, (\eta) \,/\, Z_{\Delta,K}\, (\eta) \,]^2 \leq$$

$$\leq \sum_{\zeta \in \Omega_{\Delta,K}} \gamma_{\Delta,K}\, (\zeta | \eta)\; [\, \sum_{x:\zeta_x = a} D_x( b, a, \zeta\eta_{S \setminus \Delta} ) \,]^2 \quad .$$

This can be rewritten as

$$(5.35) \qquad L(b)\; Z_{\Delta,L}\, (\eta) \,/\, Z_{\Delta,K}\, (\eta) \leq$$

$$\leq L(b)^{-1} \sum_{\zeta \in \Omega_{\Delta,K}} \; \sum_{x:\zeta_x = a} \gamma_{\Delta,L}\, ({}^x\zeta | \eta) \sum_{y:\zeta_y = a} D_y( b, a, \zeta\eta_{S \setminus \Delta} ) \quad .$$

Here we let ${}^x\zeta$ denote the configuration which is obtained from $\zeta$ by changing the value $a$ at $x$ into $b$ . Next we consider for fixed $x \in \Delta$ with $\zeta_x = a$ the difference

$$(5.36) \qquad \sum_{y:\zeta_y = a} D_y( b, a, \zeta\eta_{S \setminus \Delta} ) - \sum_{y:{}^x\zeta_y = a} D_y( b, a, {}^x\zeta\eta_{S \setminus \Delta} )$$

$$= D_x( b, a, \zeta\eta_{S\setminus\Delta} ) + \sum_{\substack{y \neq x \\ \zeta_y = a}} [ D_y( b, a, \zeta\eta_{S\setminus\Delta} ) - D_y( b, a, {}^x\zeta\eta_{S\setminus\Delta} ) ] .$$

This can be estimated by means of the inequalities

$$D_y( b, a, \zeta ) \leq \exp [ \Phi(y, a) - \Phi(x, a) + D(\Phi) ]$$

and

$$1 - D_y( b, a, {}^x\zeta ) / D_y( b, a, \zeta ) =$$

$$= 1 - \exp [ \sum_{A \supset \{x,y\}} \{ \Phi(A, {}^x\zeta) - \Phi(A, {}^{yx}\zeta) - \Phi(A, \zeta) + \Phi(A, {}^y\zeta) \} ]$$

$$\leq 2 \sum_{A \supset \{x,y\}} \max_{\omega \in \Omega_{A\setminus x}} | \Phi(A, b\omega) - \Phi(A, a\omega) | = D_{xy} .$$

Here we have written $\zeta$ instead of $\zeta\eta_{S\setminus\Delta}$, and in the last step the inequality $1 - e^s \leq |s|$ was applied. We may assume

$$C = \sup_{y \in S} [ \Phi(y, a) - \Phi(y, b) ] < \infty .$$

(Otherwise we could interchange the rôles of $a$ and $b$.) Then the r. h. s. of (5.36) has the upper bound

$$\exp [ C + D(\Phi) ] ( 1 + \sum_{y \neq x} D_{xy} )$$

$$\leq \exp [ C + D(\Phi) ] ( 1 + 2 D(\Phi) ) = M < \infty .$$

Combining this with (5.35) and again (5.28) we finally obtain

(5.37) $\quad L(b) \; Z_{\Delta,L} (\eta) / Z_{\Delta,L + 1_a - 1_b} (\eta) \leq$

$$\leq \quad M + (L(b)+1) \; Z_{\Delta,L+1_b-1_a}(\eta) \, / \, Z_{\Delta,L}(\eta) \quad .$$

Now we make the particular choice for $\Delta$ and $\eta$ which was indicated above and let $\Lambda \uparrow S$. Then $L(b)$ tends to infinity, and our final inequality gives

$$z(b, a, \omega)^{-1} \leq z(a, b, \omega) \quad . \qquad \lrcorner$$

Theorem (5.15) follows immediately from (5.31) and (5.35). Indeed, if (A) or (B) holds then any $\mu \in \text{ex}\, C$ must satisfy assertion (V) of Proposition (5.9). Hence $\mu \in G(z)$ for some $z$. Therefore the condition of Theorem (5.14) is verified.

*Bibliographical notes*: The main techniques of this paragraph appeared first in Georgii (1976 a). The estimate used in the proof of (5.34) is due to R.L. Dobrushin/R.A. Minlos (1967) and J. Ginibre (see section 3.4.9 of D. Ruelle (1969)). Under conditions which are stronger than (A), the inclusion $\text{ex}\, C \subset \bigcup_z \text{ex}\, G(z)$ follows also from Theorem 5.1 of K.G. Logan (1974), combined with Theorem (2.14). In a general setting which includes the microcanonical case C. Preston (1979) has found abstract conditions implying (5.1).

## § 6 Continuous models with interaction

We consider the continuous model which was introduced in section 1.2. In particular, we fix a diffuse, locally finite measure $\sigma$ on $\mathbb{R}^d$ and a potential $\Phi$ which satisfies the requirements of definition (1.58). We ask for conditions on $\sigma$ and $\Phi$ which imply

$$(6.1) \qquad \text{ex}\, C = \bigcup_{z \geq 0} \text{ex}\, G(z) \quad .$$

The method of attacking this problem is similar to that of § 5, but the technical details are somewhat more involved. In one respect, however, the continuous model is simpler to deal with than the discrete one: Since in the present case there is no maximal configuration, there are less symmetric functions which must be prevented from being a. s. finite. (Recall that in the case $\Phi = 0$ the necessary and sufficient condition for (6.1) in the continuous model is simpler than the corresponding one in the discrete case.) Because of this we will here be able to find sufficient conditions for (6.1) which in the case $\Phi = 0$ reduce to the necessary condition that $\sigma$ is infinite.

### 6.1 Formulation of results

In order to verify that a given measure $\mu \in \text{ex}\, C$ is a Gibbs measure with respect to an activity $z \geq 0$ it is sufficient to consider the ratios

$$(6.2) \qquad g_\Lambda^\mu (x\omega|\omega) \,/\, g_\Lambda^\mu (\omega|\omega) \qquad ( x \in \Lambda \setminus \omega \,,\; \Lambda \in S )$$

or

$$(6.3) \qquad g_\Lambda^\mu (\omega \setminus x|\omega) \,/\, g_\Lambda^\mu (\omega|\omega) \qquad ( x \in \omega_\Lambda \,,\; \Lambda \in S ) \quad .$$

Here we let

$$(6.4) \qquad g_\Lambda^\mu(\zeta|\omega) = \gamma_{\Lambda,\zeta(\Lambda)}(\zeta|\omega)\ \mu(\,N(\Lambda)=\zeta(\Lambda)|F_{S\setminus\Lambda})(\omega) \quad (\,\zeta,\omega\in\Omega\,)$$

denote the density of $\tilde{\mu}$ with respect to $\pi \otimes \mu$ on $F_\Lambda \otimes F_{S\setminus\Lambda}$. Thus by definition we have for all $A \in F_\Lambda$

$$\mu(\,A\mid F_{S\setminus\Lambda}\,) = \int_A \pi(d\zeta)\ g_\Lambda^\mu(\zeta|\,.\,) \qquad \mu\text{ - a. s.}$$

By using the approximation theorem (1.74) we will obtain explicit expressions for (6.2) and (6.3) which, up to an energy factor, are respectively given by

$$(6.5) \qquad z(\omega) = \limsup_{V\uparrow S}\ \limsup_{W\uparrow S}\ \limsup_{\Delta\uparrow S}\ z_{\Delta,W,V}(\omega)\ ,$$

$$z_{\Delta,W,V}(\omega) = Z_{\Delta\setminus W,\omega(\Delta\setminus W)-1}(\omega_{S\setminus V}) \,/\, Z_{\Delta\setminus W,\omega(\Delta\setminus W)}(\omega_{S\setminus V})$$

and

$$(6.6) \qquad z^*(\omega) = \limsup_{V\uparrow S}\ \limsup_{W\uparrow S}\ \limsup_{\Delta\uparrow S}\ z^*_{\Delta,W,V}(\omega)\ ,$$

$$z^*_{\Delta,W,V}(\omega) = Z_{\Delta\setminus W,\omega(\Delta\setminus W)+1}(\omega_{S\setminus V}) \,/\, Z_{\Delta\setminus W,\omega(\Delta\setminus W)}(\omega_{S\setminus V})\ .$$

As in (5.6) the third limiting procedure $V\uparrow S$ in performed only for the technical reason that we want to make the tail-measurability of $z(.)$ and $z^*(.)$ explicitly visible. Actually we will see that

$$z(.) = 1\,/\,z^*(.) = \lim_{\Delta\uparrow S} z_{\Delta,\emptyset,\emptyset}(.) \qquad \mu\text{ - a. s.}$$

for all $\mu \in C$ whenever (6.1) holds.

If for some $\mu$ $z(.)$ is a. s. constant then we will denote its constant value by $z(\mu)$ . $z^*(\mu)$ is defined similarly. Obviously, $z(\varepsilon_\emptyset) = z(\emptyset) = 0$ .

In section 6.2 we will prove:

(6.7) Proposition: *For any* $\mu \in \text{ex } C \setminus \{\varepsilon_\emptyset\}$ *the following statements are equivalent:*

(I) $\mu \in G(z)$ *for some* $z > 0$ .

(II) *For each* $\Lambda \in S$ *we have*

$$\mu( N(\Lambda) = 0 \mid F_{S \setminus \Lambda} ) > 0 \qquad \mu - \text{a. s.}$$

(III) *For some (all)* $\Lambda \in S$

$$\sigma \bullet \pi \bullet \mu( (x, \zeta, \omega) \in \Lambda \times \Omega^2 : g^\mu_\Lambda(x\zeta|\omega)\, g^\mu_\Lambda(\zeta|\omega) > 0 ) > 0$$

(IV) $z(\mu) > 0$

(V) $z(\mu) \vee z^*(\mu) > 0$ .

*In this case we have* $0 < z(\mu) = 1/z^*(\mu) < \infty$ , *and* $z(\mu)$ *is the unique* $z$ *such that* $\mu \in G(z)$ .

Condition (II) was introduced by Papangelou (1974) as condition $(\Sigma)$ and Kozlov (1976) as the condition of a non-degenerate vacuum and is of some interest in the theory of point processes ( see, for instance, Glötzl (1977,1978) and Kallenberg (1978) ). (Let us note that (II) is not sufficient for (I) if $\mu$ is not concentrated on the set $\Omega$ of configurations without multiple points; when multiple points are allowed it is necessary to replace statement (II) by condition $(\Sigma')$ in Matthes et al. (1977). ) Condition (III) expresses a "randomness principle" for the particle number which is similar to (5.9) (III) . Statements (IV) and (V) are the tools which enable us to find conditions on $\sigma$ and $\Phi$ which imply (6.1) .

A necessary condition can be easily found: If $\mu \in \text{ex } C \cap \bigcup_{z>0} G(z)$ then necessarily $\mu(N(S) = \infty) = 1$ . Indeed, otherwise we could find an integer $k \geq 1$ with $\mu(N(S) = k) = 1$ . Consequently, for any $\Lambda \in S$ and all integers $0 \leq l \leq k$ and $j \geq 0$ we would have

$$(6.8) \qquad \int_{\{N(S\setminus\Lambda)=l\}} \mu(d\omega) \int_{\{N(\Lambda)=j\}} \pi(d\zeta) \int_{\Lambda} \sigma(dx)\; g^{\mu}_{\Lambda}(x\zeta|\omega)\; g^{\mu}_{\Lambda}(\zeta|\omega) = 0$$

in contradiction to (6.7) (III) . Let us verify (6.8) : By the definition of $g^{\mu}_{\Lambda}$ , This integral can be written in the form

$$\int_{\Lambda} \sigma(dx) \int_{\{N(\Lambda)=j,\; N(S\setminus\Lambda)=l\}} \mu(d\omega)\; g^{\mu}_{\Lambda}(x\omega|\omega)$$

which vanishes whenever $j + l \neq k$ . Using (1.57) we find that the l. h. s. of (6.8) can also be written as

$$\int_{\{N(\Lambda)=j+1,\; N(S\setminus\Lambda)=l\}} \mu(d\omega) \int_{\Lambda} \omega(dx)\; g^{\mu}_{\Lambda}(\omega\setminus x \mid \omega)$$

which vanishes for $j + l = k$ .

From this observation we get:

(6.9) <u>Corollary:</u> *Assume*

$$\text{ex}\, C \cap \bigcup_{z>0} \text{ex}\, G(z) \neq \emptyset \quad .$$

*Then*

$$\sup_{\omega\in\Omega_{r(\Phi)}} \int \sigma(dx)\; \exp\,[-\,E(x|\omega)] = \infty \quad .$$

In particular, if $\Phi \geq 0$ then we see that $\sigma$ must be infinite. If $\Phi$ has a hard core $r(\Phi) > 0$ then it is reasonable to assume that for all $x$

$$m(x) = \inf_{\omega \in \Omega_{r(\Phi)}} E(x|\omega)$$

is finite, and (6.9) then gives the necessary condition

$$\int \sigma(dx) \exp[-m(x)] = \infty .$$

<u>Proof:</u> Suppose $\mu \in \text{ex}\, C \cap G(z)$ for some $z > 0$. Then the observation above shows that for each integer $n$ there is a set $\Lambda \in S$ with $\int \mu(d\omega)\, \omega(\Lambda) \geq n$ By formula (1.57) this gives

$$\begin{aligned} n &\leq \int \mu(d\omega) \int \pi(d\zeta)\, \gamma_\Lambda^z (\zeta|\omega)\, \zeta(\Lambda) \\ &= \int \mu(d\omega) \int_\Lambda \sigma(dx) \int \pi(d\zeta)\, \gamma_\Lambda^z (\zeta|\omega)\, z \exp[-E(x|\zeta_\Lambda\, \omega_{S\setminus\Lambda})] \\ &= z \int \mu(d\omega) \int_\Lambda \sigma(dx) \exp[-E(x|\omega)] . \end{aligned}$$

Therefore the function $\int \sigma(dx) \exp[-E(x|\cdot)]$ cannot be a.s. bounded. ⌋

Proposition (6.7) gives us some information on the structure of $C$ whenever (6.1) holds:

(6.10) <u>Theorem:</u> *Suppose*

$$\text{ex}\, C \subset \bigcup_{z \geq 0} \text{ex}\, G(z) .$$

*Then*

$$\text{ex}\, C = \bigcup_{z \geq 0} \text{ex}\, G(z) ,$$

*and the following assertions hold:*

(a) *For all* $\mu \in C$

$$E_\infty = F_\infty \qquad \mu - a.\ s.$$

(b) *For all* $\mu \in C$ *and* $\mu - a.\ a.\ \omega$

$$z(\omega) = \lim_{\Delta \uparrow S} Z_{\Delta,\omega(\Delta)-1}(\omega) \,/\, Z_{\Delta,\omega(\Delta)}(\omega) < \infty \quad .$$

(c) *For any* $\mu \in C$ , $\Lambda \in S$ *and* $\pi \otimes \mu - a.\ a.\ (\zeta, \omega)$

$$g^\mu_\Lambda (\zeta | \omega) = \gamma^{z(\omega)}_\Lambda (\zeta | \omega) \quad .$$

(d) *For each* $z \geq 0$

$$G(z) = \{ \mu \in C \ : \ z(.) = z \quad \mu - a.\ s.\ \} \quad .$$

(e) *Let* $\mu \in C$ , $m$ *the distribution of* $z(.)$ *with respect to* $\mu$ , *and* $(z, A) \to \mu_z(A)$ *a regular version of the conditional probability* $\mu( A | z(.) = z )$ $(z \geq 0,\ A \in F)$ . *Then*

$$m( z \geq 0 \ : \ \mu_z \in G(z) ) = 1$$

*and*

$$\mu = \int \mu_z \ m(dz) \quad .$$

*Moreover, if* $m_z$ *denotes the probability measure on* ex $G(z)$ *with barycentre* $\mu_z$ *then* $\int m(dz)\, m_z$ *is the probability measure on* ex $C$ *with barycentre* $\mu$ .

This theorem will be proved in section 6.2.

Now we introduce two conditions (on $\sigma$ and $\Phi$) either of which will turn out to be sufficient for (6.1) . For the sake of simplicity, we confine ourselves to pair potentials (which, of course, are supposed to satisfy the conditions of definition (1.58) ). Let $\Phi_+ (x, y) = \Phi(x, y) \vee 0$ .

(C) *$\sigma$ is infinite. $\Phi$ is bounded from below in the following sense: There is a constant* $B < \infty$ *such that*

$$(6.11) \qquad E(x|\omega) \geq -B \qquad (x \in S, \ \omega \in \Omega_{r(\Phi)}) \quad .$$

*$\Phi$ and $\sigma$ together satisfy the condition*

$$(6.12) \qquad \sup_{x \in S} \int \sigma(dy) \quad (1 - \exp[-\Phi_+(x, y)]) < \infty \quad .$$

Condition (6.11) makes sense only if $\Phi \geq 0$ or $r(\Phi) > 0$. In the latter case it is not difficult to see that (6.11) is satisfied whenever there is a monotone function $\varphi : [r(\Phi), \infty[ \to [0, \infty[$ such that $\Phi(x, y) \geq -\varphi(|x - y|)$ $(x \neq y)$ and

$$\int_{r(\Phi)}^{\infty} \varphi(s) \ s^{d-1} \ ds < \infty \quad .$$

Obviously, (6.11) is a strengthening of the stability hypothesis (1.58) (iii) and ensures that the attracting forces are not too large. On the other hand, the requirements (6.12) and $\sigma(S) = \infty$ guarantee the necessary property that for all $\mu \in \text{ex } C \setminus \{\varepsilon_\emptyset\}$ infinitely many particles exist: The repulsive forces are not too strong compared with the a priori particle rate $\sigma$.

The second condition, which we introduce now, has a similar interpretation. This condition concerns potentials without hard cores and with possibly negative values.

(D) *$\sigma$ is infinite and bounded by a multiple of Lebesgue measure, i. e.,* $\sigma \leq c\,\lambda$ *for some* $c > 0$. *$\Phi$ is bounded from below and of bounded range: There is an* $R < \infty$ *with*

$$\Phi(x, y) = 0 \qquad (|x - y| > R) \quad .$$

*$\sigma$ and $\Phi$ together satisfy the condition: For each* $a > 0$

$$(6.13) \qquad C_a = \sup_{x\in S} \int_{\{y:|y-x|>a\}} \sigma(dy)\ \Phi_+(x,y) < \infty .$$

Clearly, (6.13) implies (6.12) when $\sigma \le c\,\lambda$ . Moreover, (6.13) holds only if $r(\Phi) = 0$ .

Observe that the essence of both the conditions (C) and (D) is the spatial uniformity. If $\sigma = \lambda$ and $\Phi$ is shift-invariant then (C) and (D) reduce to rather mild regularity assumptions.

(6.14) Theorem: *Suppose* $\sigma$ *and* $\Phi$ *satisfy one of the conditions* (C) *and* (D) *Then*

$$\text{ex } C = \bigcup_{z\ge 0} \text{ex } G(z) ,$$

*and the statements* (a) - (e) *of* (6.10) *hold.*

We prove this in section 6.3 .

## 6.2 Conditional densities and the activity functions

We fix a measure $\mu \in \text{ex } C \setminus \{\varepsilon_\emptyset\}$ and consider the densities

$$g_\Lambda(\cdot|\cdot) = d\tilde{\mu} / d(\pi \otimes \mu) \mid F_\Lambda \otimes F_{S\setminus\Lambda}$$

which were introduced in (6.4) . By definition $g_\Lambda(\omega|\omega) > 0$ for $\mu$ - a. a. $\omega$

For an arbitrary probability measure $\nu$ on $(\Omega, F)$ we let $C_\nu$ denote the corresponding Campbell measure on $(S \times \Omega,\ \mathcal{B} \otimes F)$ which is defined by

$$(6.15) \qquad C_\nu (B \times A) = \int_A \nu(d\omega)\ \omega(B) \qquad (A \in F,\ B \in \mathcal{B}) .$$

Our first aim is to use the approximation theorem (1.74) in order to express the ratios (6.2) and (6.3) by means of the densities $\gamma_{W\Delta,\omega(\Delta)}(\zeta|\omega)$ (which are defined in (1.73) ).

(6.16) Lemma: *Let* $\Lambda \in S$ . *Then for* $\sigma \otimes \mu$ *- a. a.* $(x, \omega) \in \Lambda \times \Omega$

$$g_\Lambda(x\omega|\omega) \;/\; g_\Lambda(\omega|\omega) =$$

$$= \lim_{W\uparrow S} \lim_{\Delta\uparrow S} \gamma_{W\Delta,\omega(\Delta)}(x\omega|\omega) \;/\; \gamma_{W\Delta,\omega(\Delta)}(\omega|\omega) ,$$

*and for* $C_\mu$ *- a. a.* $(x, \omega) \in \Lambda \times \Omega$

$$g_\Lambda(\omega \setminus x|\omega) \;/\; g_\Lambda(\omega|\omega) =$$

$$= \lim_{W\uparrow S} \lim_{\Delta\uparrow S} \gamma_{W\Delta,\omega(\Delta)}(\omega \setminus x|\omega) \;/\; \gamma_{W\Delta,\omega(\Delta)}(\omega|\omega) .$$

Proof: 1. For any $W \in S$ let $u_W$ denote the density of $\mu|F_W$ with respect to $\pi|F_W$ . Then for $\pi \otimes \mu$ - a. a. $(\zeta, \omega)$ we have

(6.17) $$g_\Lambda(\zeta|\omega) = \lim_{W\uparrow S} u_W(\zeta_\Lambda\, \omega_{S\setminus\Lambda}) \;/\; u_{W\setminus\Lambda}(\omega) .$$

This follows from the martingale convergence theorem. Indeed, $F_\Lambda \otimes F_{S\setminus\Lambda}$ is generated by the algebra $\bigcup_W F_\Lambda \otimes F_{W\setminus\Lambda}$ , and it is easily verified that

$$(\zeta, \omega) \to u_W(\zeta_\Lambda\, \omega_{S\setminus\Lambda}) \;/\; u_{W\setminus\Lambda}(\omega)$$

is $\pi \otimes \mu$ - a. s. well-defined and a density of $\tilde{\mu}$ with respect to $\pi \otimes \mu$ on $F_\Lambda \otimes F_{W\setminus\Lambda}$ .

2. For $\sigma \otimes \pi \otimes \mu$ - a. a. $(x, \zeta, \omega) \in \Lambda \times \Omega \times \Omega$

$$g_\Lambda(x\zeta|\omega) = \lim_{W\uparrow S} u_W(x\zeta_\Lambda\, \omega_{S\setminus\Lambda}) \;/\; u_{W\setminus\Lambda}(\omega)$$

and for $C_\pi \otimes \mu$ - a. a. $(x, \zeta, \omega) \in \Lambda \times \Omega \times \Omega$

$$g_\Lambda(\zeta \setminus x|\omega) \quad = \quad \lim_{W \uparrow S} \quad u_W(\zeta_\Lambda\, \omega_{S \setminus \Lambda} \setminus x\,) \,/\, u_{W \setminus \Lambda}(\omega) \quad .$$

In order to see this let $h$ denote the indicator function of the set of all $(\zeta, \omega)$ for which (6.17) does not hold. Then steps 1. and (1.57) give

$$\int_\Lambda \sigma(dx) \quad \int \pi(d\zeta) \quad \int \mu(d\omega) \quad h(x\zeta, \omega)$$

$$= \quad \int \pi(d\zeta) \quad \int \mu(d\omega) \quad h(\zeta, \omega) \quad \zeta(\Lambda) \quad = \quad 0 \quad .$$

This proves the first assertion. The second follows from

$$\int \pi(d\zeta) \quad \int_\Lambda \zeta(dx) \quad \int \mu(d\omega) \quad h(\zeta \setminus x, \omega) \quad =$$

$$= \quad \int_\Lambda \sigma(dx) \quad \int \pi(d\zeta) \quad \int \mu(d\omega) \quad h(\zeta, \omega) \quad = \quad 0 \quad .$$

3. Suppose $\Lambda \subset W \in \mathcal{S}$ . Then for $\pi \otimes \mu$ - a. a. $(\zeta, \omega)$

$$u_W(\,\zeta_\Lambda\, \omega_{S \setminus \Lambda}\,) \quad = \quad \lim_{\Delta \uparrow S} \quad \gamma_{W\Delta,\omega(\Delta)} \quad (\,\zeta_\Lambda\, \omega_{S \setminus \Lambda}|\omega\,) \quad .$$

Indeed, from (1.74) and (1.56) we obtain

$$u_W(\,\zeta_\Lambda\, \eta_{S \setminus \Lambda}\,) \quad = \quad \lim_{\Delta \uparrow S} \quad \gamma_{W\Delta,\omega(\Delta)} \quad (\,\zeta_\Lambda\, \eta_{S \setminus \Lambda}|\omega\,)$$

for $\pi \otimes \pi \otimes \mu$ - a. a. $(\zeta, \eta, \omega)$ . The indicator function $h$ of the set of all $(\zeta, \eta, \omega)$ for which this assertion fails is measurable with respect to $F_\Lambda \otimes F_{W \setminus \Lambda} \otimes E_W$ . Thus

$$0 = \int \pi(d\zeta) \int \pi(d\eta) \int \mu(d\omega)\ \gamma_{W,\omega(W)}\ (\eta|\omega)\ h(\zeta, \eta, \omega)$$

$$= \int \pi(d\zeta) \int \mu(d\omega)\ h(\zeta, \omega, \omega) \quad .$$

4. After these preparations we are now ready to prove the lemma. The steps 1. and 2. imply that for $\sigma \otimes \pi \otimes \mu$ - a. a. $(x, \zeta, \omega) \in \Lambda \times \{g_\Lambda > 0\}$

(6.18) $$g_\Lambda(x\zeta|\omega) \,/\, g_\Lambda(\zeta|\omega) =$$

$$= \lim_{W \uparrow S} u_W(\, x\zeta_\Lambda\ \omega_{S\setminus\Lambda}\,) \,/\, u_W(\, \zeta_\Lambda\ \omega_{S\setminus\Lambda}\,) \quad .$$

Next we apply the result of 3. together with the identity which is obtained from 3. when $\zeta$ is replaced by $x\zeta$ with $x$ distributed according to $\sigma|_\Lambda$ . (This statement can be proved by an argument similar to that in 2. ). Inserting these identities into (6.18) we see that the l. h. s. of (6.18) can be expressed as a twofold limit of ratios of $\gamma$'s . By absolute continuity, this identity is true for $\sigma \otimes \tilde{\mu}$ - a. a. $(x, \zeta, \omega) \in \Lambda \times \{g_\Lambda > 0\}$ . Since $\tilde{\mu}$ is supported on $\{g_\Lambda > 0\}$ and on the diagonal of $\Omega \times \Omega$ , the first assertion follows. The second assertion is obtained similarly: According to 1. - 3. , for $C_\pi \otimes \mu$ - a. a. $(x, \zeta, \omega) \in \Lambda \times \{g_\Lambda > 0\}$ we have an identity which corresponds to (6.18) and a similar identity involving the twofold limit of **ratios** of $\gamma$'s . Denoting by $h$ the indicator function of all $(x, \zeta, \omega)$ for which this identity is violated, we have

$$0 = \int \pi(d\zeta) \int_\Lambda \zeta(dx) \int \mu(d\omega)\ \gamma_{\Lambda,\omega(\Lambda)}\ (\zeta|\omega)\ h(x, \zeta, \omega)$$

$$= \int \mu(d\omega) \int_\Lambda \omega(dx)\ h(x, \omega, \omega) \quad .$$

This proves the second assertion. $\lrcorner$

(6.19) Proposition: *Let* $\Lambda \in S$ . *Then for* $\sigma \otimes \pi \otimes \mu$ - a. a. $(x, \zeta, \omega)$ $\in \Lambda \times \Omega \times \Omega$ *we have*

(i) $$g_\Lambda(x\zeta|\omega) = z(\mu)\ \exp[\ - E(\ x|\zeta_\Lambda\ \omega_{S\smallsetminus\Lambda}\ )\ ]\ \ g_\Lambda(\zeta|\omega)$$

*provided that* $g_\Lambda(\zeta|\omega) > 0$

(ii) $$g_\Lambda(\zeta|\omega) = z^*(\mu)\ \exp[\ E(\ x|\zeta_\Lambda\ \omega_{S\smallsetminus\Lambda}\ )\ ]\ \ g_\Lambda(x\zeta|\omega)$$

*provided that* $g_\Lambda(x\zeta|\omega) > 0$ .

Proof: "(i)": Let $\Lambda \subset V \subset W \subset \Delta \in S$ . For $\mu$ - a. a. $\omega$ and all $x \in \Lambda \smallsetminus \omega$ we may write

$$\gamma_{W\Delta,\omega(\Delta)}(x\omega|\omega)\ /\ \gamma_{W\Delta,\omega(\Delta)}(\omega|\omega) =$$

$$= \frac{\displaystyle\int_{\{N(\Delta\smallsetminus W)=\omega(\Delta\smallsetminus W)-1\}} \pi(d\alpha)\ \exp[\ - E_\Delta(\ x\omega_W\ \alpha_{\Delta\smallsetminus W}|\omega\ )\ ]}{\displaystyle\int_{\{N(\Delta\smallsetminus W)=\omega(\Delta\smallsetminus W)\}} \pi(d\alpha)\ \exp[\ - E_\Delta(\ \omega_W\ \alpha_{\Delta\smallsetminus W}|\omega\ )\ ]} .$$

Now we observe that

$$E_\Delta(\ x\omega_W\ \alpha_{\Delta\smallsetminus W}|\omega\ ) = E(\ x|\omega_W\ \alpha_{\Delta\smallsetminus W}\ \omega_{S\smallsetminus\Delta}\ ) + E_\Delta(\ \omega_W\ \alpha_{\Delta\smallsetminus W}|\omega\ )$$

and

$$E_\Delta(\ \omega_W\ \alpha_{\Delta\smallsetminus W}|\omega\ ) = E_{\Delta\smallsetminus W}(\alpha|\omega_{S\smallsetminus V}) + E_V(\ \omega|\omega_W\ \alpha_{\Delta\smallsetminus W}\ \omega_{S\smallsetminus\Delta}\ ) + K$$

whenever $\omega_W\ \alpha_{\Delta\smallsetminus W}\ \omega_{S\smallsetminus\Delta} \in \Omega_{r(\Phi)}$ ; here $K$ denotes an energy term which depends on $\omega$ but not on $\alpha$ . The regularity hypothesis (1.58) (iv) implies that

$$\sup\{\ |\ E(x|\omega) - E(x|\alpha)\ |\ :\ \alpha \in \Omega_{r(\Phi)}\ ,\ \alpha_W = \omega_W\ \}$$

and

$$\sup\{\ |\ E_V(\omega|\omega) - E_V(\omega|\alpha)\ |\ :\ \alpha \in \Omega_{r(\Phi)},\ \alpha_W = \omega_W\ \}$$

vanish in the limit $W \uparrow S$. By an argument similar to the one used in the proof of (5.17) we obtain from (6.16) that

$$g_\Lambda(x\omega|\omega) \,/\, g_\Lambda(\omega|\omega) \quad = \quad z(\mu) \;\exp\,[\, - E(x|\omega)\,]$$

for $\sigma \otimes \mu$ - a. a. $(x, \omega) \in \Lambda \times \Omega$. This is equivalent to assertion (i) since $\tilde{\mu}$ is equivalent to $\pi \otimes \mu$ on $(F_\Lambda \otimes F_{S\setminus\Lambda}) \cap \{g_\Lambda > 0\}$.

"(ii)" The same argument as above shows that

$$g_\Lambda(\omega \setminus x|\omega) \,/\, g_\Lambda(\omega|\omega) \quad = \quad z^*(\mu) \;\exp E(x|\omega)$$

for $C_\mu$ - a. a. $(x, \omega) \in \Lambda \times \Omega$. Therefore we have for the indicator function $h$ of the set of all $(x, \zeta, \omega)$ for which assertion (ii) fails to hold

$$\begin{aligned} 0 &= \int \mu(d\omega) \int_\Lambda \omega(dx) \quad h(x, \omega \setminus x, \omega) \\ &= \int \pi(d\zeta) \int \mu(d\omega) \;\; g_\Lambda(\zeta|\omega) \int_\Lambda \zeta(dx) \;\; h(x, \zeta \setminus x, \omega) \\ &= \int_\Lambda \sigma(dx) \int \pi(d\zeta) \int \mu(d\omega) \;\; g_\Lambda(x\zeta|\omega) \;\; h(x, \zeta, \omega) \; . \end{aligned}$$

┘

Now we are prepared to start the proof of proposition (6.7). The implication (I) ⇒ (II) is trivial since $\pi(N(\Lambda) = 0) > 0$ and $\gamma^z_\Lambda(\emptyset|.) > 0$ whenever $\Lambda \in S$ and $z \geq 0$. The step (II) ⇒ (III) is accomplished by the next

(6.20) <u>Lemma:</u> *Suppose that*

$$\mu(\; N(\Lambda) = 0 \mid F_{S\setminus\Lambda} \;) \;>\; 0 \qquad \mu \text{ - a. s.}$$

*for all* $\Lambda \in S$. *Then*

(6.21) $\quad \sigma \otimes \pi \otimes \mu(\; (x, \zeta, \omega) \in \Lambda \times \Omega \times \Omega \;:\; g_\Lambda(x\zeta|\omega) \;\; g_\Lambda(\zeta|\omega) \;>0\;) \;>\; 0$

*for every* $\Lambda \in S$ .

Proof: 1. First we prove that for all $\Lambda \in S$ the measure

$$\nu_\Lambda \ : \ A \ \to \int \mu(d\omega) \int_\Lambda \omega(dx) \quad 1_A \ (\omega \setminus x)$$

on F is absolutely continuous with respect to $\mu$ . As Kallenberg (1978) has shown, this follows from our condition and the fact that $\mu$ a. s. has no multiple points. We reproduce his argument: Let $A \in F$ such that $\mu(A) = 0$ . Then for all bounded $\Delta \in B$ we have

$$\nu_\Delta( \ A \cap \{N(\Delta) = 0\} \ ) \quad = \quad 0 \quad .$$

Indeed, from the identity

$$\int_{\{\omega : \omega_{S \setminus \Delta} \in A\}} \mu(N(\Delta)=0 \mid F_{S \setminus \Delta}) \ d\mu \quad = \quad \mu(A \cap \{N(\Delta)=0\}) = 0$$

and the hypothesis $\mu( \ N(\Delta) = 0 \mid F_{S \setminus \Delta} \ ) \ > \ 0$ a. s. we get

$$\mu( \ \omega \ : \ \omega_{S \setminus \Delta} \in A \ ) \quad = \quad 0 \quad .$$

Hence

$$\nu_\Delta( \ A \cap \{N(\Delta) = 0\} \ ) \quad =$$

$$= \int \mu(d\omega) \int_\Delta \omega(dx) \ \ 1_{\{N(\Delta)=0\}} \ (\omega \setminus x) \ \ 1_A \ (\omega_{S \setminus \Delta}) \ = \ 0 \quad .$$

Now let $\Lambda \in S$ . We may assume

$$A \subset \{ \ N(\Lambda) \ < \ N \ \}$$

for some integer $N$ . (For if $\nu_\Lambda( A \cap \{N(\Lambda) < N\} ) = 0$ for all $N$ then $\nu_\Lambda(A) = 0$ . ) We choose a sequence $(P_n)_{n\geq 1}$ of finite measurable partitions of $\Lambda$ obtained by successive refinements such that $\max\{ \operatorname{diam} \Delta : \Delta \in P_n \}$ tends to zero when $n \to \infty$ . Then we can write

$$\nu_\Lambda(A) = \sum_{\Delta\in P_n} \nu_\Delta( A \cap \{N(\Delta) \geq 1\} )$$

$$= \sum_{\Delta\in P_n} \int_{\{N(\Delta)\geq 2\}} \mu(d\omega) \int_\Delta \omega(dx)\ 1_A\ (\omega \setminus x)$$

$$\leq N \sum_{\Delta\in P_n} \mu(N(\Delta) \geq 2\ ,\ N(\Lambda) \leq N)$$

$$= N \int_{\{N(\Lambda)\leq N\}} \sum_{\Delta\in P_n} I_{\{N(\Delta) \geq 2\}}\ d\mu .$$

By the dominated convergence theorem, this tends to zero when $n$ tends to infinity.

2. Now we prove (6.21). We choose a set $\Lambda \in S$ and show

$$\int \mu(d\omega) \int_\Lambda \sigma(dx) \int \pi(d\zeta)\ g_\Lambda\ (x\zeta|\omega)\ g_\Lambda\ (\zeta|\omega) > 0 .$$

This integral can be written as

$$\int \mu(d\omega) \int \pi(d\zeta) \int_\Lambda \zeta(dx)\ g_\Lambda(\zeta|\omega)\ g_\Lambda\ (\zeta\setminus x|\omega)$$

$$= \int \mu(d\omega) \int_\Lambda \omega(dx)\ g_\Lambda(\omega\setminus x|\omega)$$

$$= \int \nu_\Lambda(d\omega)\ g_\Lambda(\omega|\omega) .$$

This expression is positive. Indeed,

$$\mu(\omega : g_\Lambda(\omega|\omega) = 0) = \int \pi(d\zeta) \int \mu(d\omega)\ g_\Lambda(\zeta|\omega)\ 1_{\{g_\Lambda = 0\}}\ (\zeta,\omega) = 0$$

and thus by 1.

$$\nu_\Lambda(\omega : g_\Lambda(\omega|\omega) = 0) = 0 .$$

Moreover,

$$\nu_\Lambda(\Omega) = \int \mu(d\omega)\ \omega(\Lambda) > 0$$

because for sufficiently large $\Delta \supset \Lambda$

$$\mu(N(\Delta) > 0) > 0$$

and therefore

$$\int \mu(d\omega)\ \omega(\Lambda) = \int \mu(d\omega) \int \pi(d\zeta)\ \zeta(\Lambda)\ \gamma_{\Delta,\omega(\Delta)}(\zeta|\omega) > 0 . \qquad \lrcorner$$

Next we prove the equivalence of the statements (6.7) (III) - (V):

(6.22) Lemma: *If (6.21) holds for some* $\Lambda \in S$ *then* $z(\mu) > 0$. *Conversely, if* $z(\mu) \vee z^*(\mu) > 0$ *then (6.21) is satisfied for all* $\Lambda \in S$.

Proof: 1. If $z(\mu) = 0$ then (6.19) (i) shows that for each $\Lambda \in S$ (6.21) fails.

2. Suppose $z(\mu) > 0$. Then again (6.19) (i) implies that (6.21) is true for all $\Lambda \in S$ except for those with

$$\sigma \otimes \pi \otimes \mu\ ((x,\zeta,\omega) \in \Lambda \times \Omega \times \Omega : g_\Lambda(\zeta|\omega) > 0,\ x\,\zeta_\Lambda\,\omega_{S\setminus\Lambda} \in \Omega_r) = 0 ;$$

here $r = r(\Phi)$. Clearly there is no such exception when $r = 0$. If $r > 0$ then we have to use the hypothesis that $\mu$ is supported on $\Omega_r$. This hypothesis guarantees the existence of a set $\Delta \in S$ with $\Delta \supset \Lambda$ and

$$\mu(N(\Delta) < M\ ) > 0 .$$

For all $\omega \in \Omega_r$ with $\omega(\Delta) < M_\Delta$ we have

$$\pi(\ \zeta \in \Omega : \zeta(\Delta) = \omega(\Delta) + 1,\ \zeta_\Delta\ \omega_{S\smallsetminus\Delta} \in \Omega_r\ ) > 0\ .$$

Hence

$$\int \mu(d\omega) \int \pi(d\zeta)\ 1_{\Omega_r}(\zeta_\Delta\omega_{S\smallsetminus\Delta}) \int_\Lambda \zeta(dx)\ \gamma_{\Delta,\omega(\Delta)}(\zeta\smallsetminus x|\omega) > 0\ .$$

But this integral can be written as

$$\int_\Lambda \sigma(dx) \int \pi(d\zeta) \int \mu(d\omega)\ \gamma_{\Delta,\omega(\Delta)}(\zeta|\omega)\ \ 1_{\Omega_r}(x\zeta_\Delta\omega_{S\smallsetminus\Delta})$$

$$= \int_\Lambda \sigma(dx) \int \mu(d\omega)\ \ 1_{\Omega_r}(x\omega)$$

$$= \int_\Lambda \sigma(dx) \int \pi(d\zeta) \int \mu(d\omega)\ g_\Lambda(\zeta|\omega)\ 1_{\Omega_r}(\zeta_\Lambda\omega_{S\smallsetminus\Lambda})\ .$$

This shows that also in the case $r > 0$ there are no exceptional $\Lambda$'s .

3. Finally we assume $z^*(\mu) > 0$ . For all $\Lambda \in S$ we have ( cf. the end of the proof of (6.20) )

$$0 < \int \mu(d\omega)\ \omega(\Lambda) = \int \pi(d\zeta) \int \mu(d\omega)\ g_\Lambda(\zeta|\omega)\ \zeta(\Lambda)$$

$$= \int_\Lambda \sigma(dx) \int \pi(d\zeta) \int \mu(d\omega)\ g_\Lambda(x\zeta|\omega)\ .$$

Thus (6.19) (ii) implies (6.21) for all $\Lambda \in S$ . ┘

The last step in the proof of (6.7) is provided by

(6.23) Lemma: *Suppose* $z(\mu) \vee z^*(\mu) > 0$ . *Then* $0 < z(\mu) = 1/z^*(\mu) < \infty$ *and* $\mu \in G(z(\mu))$ .

Proof: From (6.22) and (6.19) we know that for any $\Lambda \in S$ the set of all $(x,\zeta,\omega) \in \Lambda \times \Omega \times \Omega$ with

$$(6.24) \qquad g_\Lambda(x\zeta|\omega) = z(\mu)\ \exp[-E(x|\zeta_\Lambda\ \omega_{S\setminus\Lambda})]\ g_\Lambda(\zeta|\omega)$$

$$(6.25) \qquad g_\Lambda(\zeta|\omega) = z^*(\mu)\ \exp[E(x|\zeta_\Lambda\ \omega_{S\setminus\Lambda})]\ g_\Lambda(x\zeta|\omega)$$

and

$$g_\Lambda(x\zeta|\omega) > 0,\ g_\Lambda(\zeta|\omega) > 0$$

has positive measure with respect to $\sigma \otimes \pi \otimes \mu$. For these $(x,\zeta,\omega)$ $E(x|\zeta_\Lambda\ \omega_{S\setminus\Lambda})$ is necessarily finite whenever $z(\mu) \vee z^*(\mu) > 0$, and this implies $z(\mu)\ z^*(\mu) > 0$ and $z(\mu) = 1/z^*(\mu)$.

Now again (6.19) tells us that a.s. with respect to $\sigma \otimes \pi \otimes \mu$ the following holds: If either $g_\Lambda(\zeta|\omega) > 0$ or $g_\Lambda(x\zeta|\omega) > 0$ then $g_\Lambda(\zeta|\omega) > 0$ and (6.24) is satisfied. Thus (6.24) holds almost surely.

This in turn implies

$$(6.26) \qquad g_\Lambda(\zeta|\omega) = z(\mu)^{\zeta(\Lambda)}\ \exp[-E_\Lambda(\zeta|\omega)]\ g_\Lambda(\emptyset|\omega)$$

for $\pi \otimes \mu$-a.a. $(\zeta,\omega)$. Indeed, letting $d_\Lambda(\zeta|\omega)$ denote the difference between the two sides of (6.26) we clearly have $d_\Lambda(\emptyset|\omega) = 0$ a.s. . It follows by induction on $N \geq 0$ that $d_\Lambda(.|.) = 0$ a.s. on $\{N(\Lambda)=N\}\times\Omega$ . This can be seen from the equation

$$(N+1) \int_{\{N(\Lambda)=N+1\}} \pi(d\zeta) \int \mu(d\omega)\ |\ d_\Lambda(\zeta|\omega)\ |$$

$$= \int_\Lambda \sigma(dx) \int \pi(d\zeta) \int \mu(d\omega)\ |\ d_\Lambda(x\zeta|\omega)\ |\ 1_{\{N(\Lambda)=N+1\}}(x\zeta)$$

$$= \int_\Lambda \sigma(dx) \int_{\{N(\Lambda)=N\}} \pi(d\zeta) \int \mu(d\omega)\ |\ d_\Lambda(\zeta|\omega)\ |\ z(\mu)\ \exp[-E(x|\zeta_\Lambda\ \omega_{S\setminus\Lambda})].$$

Since (6.26) is equivalent to

$$g_\Lambda(\cdot|\cdot) = \gamma_\Lambda^{z(\mu)}(\cdot|\cdot) \quad \pi \otimes \mu\text{-a.s.},$$

the proof of the lemma is complete. $\lrcorner$

We finish the proof of proposition (6.7) by noting that if $\mu \in \text{ex}\, C \cap G(z)$ for some $z > 0$ then by (6.19) (i) we have $z = z(\mu)$.

Now we proceed to the

(6.27) Proof of Theorem (6.10):

1. If $\mu \in C$ then $\mu(z(\cdot) < \infty) = 1$. Indeed, (6.7) asserts that this is true for all $\mu \in \text{ex}\, C = \text{ex}\, C \cap \bigcup_{z \geq 0} G(z)$.

2. The proof of the second assertion in (b) is similar to that of the corresponding result in (6.14) (b). Let $\mu \in \text{ex}\, C$. Then by assumption $\mu \in G(z)$ for $z = z(\mu)$. We form the disjoint union $\underline{S} = S \cup K$ of $S$ and a copy of the unit cube $K \subset \mathbf{R}^d$ and denote by $\underline{\sigma}$ the measure on $\underline{S}$ which equals $\sigma$ on $S$ and Lebesgue measure on $K$. Furthermore, we define a potential $\underline{\Phi}$ on $\underline{S}$ by

$$\underline{\Phi}(\underline{\alpha}) = \begin{cases} \Phi(\underline{\alpha}) & \text{if } \underline{\alpha} \subset S \\ 0 & \text{otherwise} \end{cases}$$

and denote by $\underline{G}(z)$ and $\underline{C}$ the corresponding sets of Gibbs measures and canonical Gibbs measures on $\underline{\Omega} = \Omega_K \times \Omega$, where $\Omega_K$ denotes the set of all finite subsets of $K$. We put

$$\underline{\mu} = \pi_K^z \otimes \mu .$$

Here $\pi_K^z$ denotes the Poisson process on $\Omega_K$ with intensity measure $z\lambda|_K$. It is not difficult to verify $\underline{\mu} \in \underline{G}(z) \subset \underline{C}$. Moreover, $\underline{\mu}$ is trivial on $\underline{E}_\infty$. In order to see this let $\underline{A} \in \underline{E}_\infty$ be such that $\underline{\mu}(\underline{A}) > 0$. Then $\underline{\nu} = \underline{\mu}(\cdot|\underline{A}) \in \underline{C}$ and $\underline{\nu}(N(S) = \infty) > 0$ since, because of the observation preceding (6.9), $\mu(N(S) = \infty) = 1$. Thus for any $N \geq 0$ there is a set $\underline{\Delta} \in \underline{S}$ such that $\underline{\Delta} \supset K$ and $\underline{\nu}(N(\underline{\Delta}) \geq N) > 0$.

This implies

$$\underline{\nu}(\ N(K)=N\ ) \geq \int_{\{N(\underline{\Delta})\geq N\}} \underline{\nu}(d\underline{\omega}) \int_{\{N(K)=N\}} \underline{\pi}(d\underline{\zeta})\ \underline{\gamma}_{\underline{\Delta},\underline{\omega}(\underline{\Delta})}(\underline{\zeta}|\underline{\omega}) > 0$$

for all $N \geq 0$ . Hence for any $N \geq 0$ we have $\underline{\mu}(\{N(K)=N\}\cap\underline{A}) > 0$ and thus $\mu(\ A_N\ ) > 0$ with $A_N \in E_\infty$ denoting the set for which $\{N(K)=N\} \cap A_N = \{N(K)=N\} \cap \underline{A}$ . Since $\mu$ is trivial on $E_\infty$ , $\mu(\ A_N\ ) = 1$ for all $N$ and therefore

$$\underline{\mu}(\ \underline{A}\ ) = \sum_{N\geq 0} \pi_K^z(\ N(K)=N\ )\ \mu(\ A_N\ ) = 1 \quad ,$$

as desired. Now we apply the approximation theorem (1.74), which gives for $\lambda \otimes \mu$-a.a. $(x,\omega) \in K \times \Omega$

$$z = \underline{u}_K(\{x\})\ /\ \underline{u}_K(\emptyset) = \lim_{\underline{\Lambda}\uparrow\underline{S}} \underline{\gamma}_{K\underline{\Lambda},\omega(\underline{\Lambda})}(\{x\}|\omega)\ /\ \underline{\gamma}_{K\underline{\Lambda},\omega(\underline{\Lambda})}(\emptyset|\omega)$$

$$= \lim_{\Lambda\uparrow S} Z_{\Lambda,\omega(\Lambda)-1}(\omega)\ /\ Z_{\Lambda,\omega(\Lambda)}(\omega) \quad .$$

3. The proofs of statements (c), (a), (d) are completely analogous to those of (5.14) (c), (a), (d) . Moreover, (a) implies

$$\bigcup_{z\geq 0} \text{ex } G(z) \subset \text{ex } C$$

because of the characterization of ex $G(z)$ and ex $C$ in terms of zero-one laws for $F_\infty$ and $E_\infty$ respectively.

4. In the proof of (5.14) (e) we have used the fact that $C$ is weakly compact in the discrete case. In order to prove (6.10)(e) we will use another argument which is actually more elementary (and works also in the discrete case). Let $\Lambda \in S$ , $A \in F_\Lambda$, $B \in F_{S\setminus\Lambda}$, and $C = \{z(.) \in C'\} \in F_\infty$ for some Borel set $C' \subset [0,\infty[$ .

Then from (c) we obtain

$$\int_{C'} m(dz)\ \mu_z(A \cap B) = \int_{C} d\mu\ \mu(A \cap B | z(\cdot))$$

$$= \quad \mu(A \cap B \cap C) = \int_A \pi(d\zeta) \int_{B\cap C} \mu(d\omega)\ \gamma_\Lambda^{z(\omega)}(\zeta|\omega)$$

$$= \quad \int_{C'} m(dz) \int_A \pi(d\zeta) \int_B \mu_z(d\omega)\ \gamma_\Lambda^{z}\ (\zeta|\omega)\ .$$

Since $F_\Lambda$ and $F_{S\setminus\Lambda}$ are countably generated we see that for m-a.a. z

$$d\tilde{\mu}_z \ /\ d(\pi \otimes \mu_z)\ |\ F_\Lambda \otimes F_{S\setminus\Lambda} = \gamma_\Lambda^{z}(\cdot|\cdot) \text{ a.s.}\ .$$

Letting $\Lambda$ run through a cofinal sequence in $S$ we conclude

$$m(\ z\geq 0\ :\ \ \mu_z \in G(z)\ ) \ = \ 1 \ .$$

The other statements in (e) are evident. ⌟

## 6.3 Estimating the activity functions

Here we use conditions (C) and (D) to estimate the functions $z(\cdot)$ and $z^*(.)$ which were defined by (6.5) and (6.6). These estimates are based on the following continuous counterpart to (5.28):

(6.38) Remark: *Suppose* $\Lambda \in S$, $N \geq 0$ *and* $\omega \in \Omega_{(r(\Phi))}$ *with* $Z_{\Lambda,N}(\omega) > 0$. *Then*

$$(N+1)\ Z_{\Lambda,N+1}(\omega)\ /\ Z_{\Lambda,N}(\omega) = \int \pi(d\zeta)\ \gamma_{\Lambda,N}(\zeta|\omega) \int \sigma(dx)\ \exp[-E(x|\zeta_\Lambda\ \omega_{S\setminus\Lambda})]\ .$$

Proof: The ratio on the l.h.s. coincides with

$$Z_{\Lambda,N}(\omega)^{-1} \int_{\{N(\Lambda)=N+1\}} \pi(d\zeta)\ \zeta(\Lambda)\ \exp[-E_\Lambda(\zeta|\omega)]$$

$$= \quad Z_{\Lambda,N}(\omega)^{-1} \int_\Lambda \sigma(dx) \int_{\{N(\Lambda)=N\}} \pi(d\zeta)\ \exp[-E_\Lambda(x\zeta|\omega)]$$

which equals the r.h.s. ⌟

(6.29) Lemma: *Assume* (C) *or* (D). *Then*

$$\mu(N(S) = \infty) = 1$$

*for all* $\mu \in \text{ex}\ C \setminus \{\varepsilon_\emptyset\}$.

Proof: Let $\mu \in \text{ex}\ C$ and suppose $\mu(N(S) < \infty) > 0$. Then there is an integer $N \geq 1$ such that $\mu(N(S) = N) = 1$. Because of (1.74) this implies that for all $\Lambda \in S$

(6.30) $$u_\Lambda = \lim_{\Delta \uparrow S} \gamma_{\Lambda\Delta,N}(.|\emptyset) \qquad \pi\text{-a.s.}$$

where $u_\Lambda = d\mu / d\pi \mid F_\Lambda$. We prove that this is impossible except when $\mu = \varepsilon_\emptyset$

Clearly, (6.30) shows

$$u_\Lambda = 0 \quad \pi\text{-a.s.} \quad \text{on} \quad \{N(\Lambda) > N\}.$$

Moreover, for $\sigma \otimes \pi$-a.a. $(x,\zeta) \in \Lambda \times \Omega$ we get

(6.31) $$u_\Lambda(\zeta) / u_\Lambda(x\zeta) = \lim_{\Delta \uparrow S} \gamma_{\Lambda\Delta,N}(\zeta|\emptyset) / \gamma_{\Lambda\Delta,N}(x\zeta|\emptyset)$$

$$= \lim_{\Delta\uparrow S} \frac{1}{n+1} \int_{\{N(\Delta\setminus\Lambda)=n\}} \pi(d\alpha) \int_{\Delta\setminus\Lambda} \sigma(dy) \exp[-E_\Delta(y\zeta_\Lambda\alpha_{\Delta\setminus\Lambda}|\emptyset)] \Big/$$

$$\Big/ \int_{\{N(\Delta\setminus\Lambda)=n\}} \pi(d\alpha) \exp[-E_\Delta(x\zeta_\Lambda\alpha_{\Delta\setminus\Lambda}|\emptyset)]$$

whenever $u_\Lambda(x\zeta) > 0$ and $n = N - \zeta(\Lambda)-1 \geq 0$. We will show that either of the conditions (C) and (D) implies that the r.h.s. of (6.31) is infinite. This will prove $\mu = \varepsilon_\emptyset$. Indeed, it shows that $u_\Lambda(x\zeta) = 0$ for $\sigma \otimes \pi$-a.a. $(x,\zeta) \in \Lambda \times \Omega$ with $\zeta(\Lambda) \leq N-1$ and therefore

$$\int \mu(d\omega)\, \omega(\Lambda) = \int_{\{N(\Lambda)\leq N\}} \pi(d\zeta)\, \zeta(\Lambda)\, u_\Lambda(\zeta)$$

$$= \int_\Lambda \sigma(dx) \int_{\{N(\Lambda)\leq N-1\}} \pi(d\zeta)\, u_\Lambda(x\zeta) = 0 .$$

Since $\Lambda$ is arbitrary we get the desired conclusion $\mu = \varepsilon_\emptyset$.

In order to estimate the r.h.s. of (6.31) we observe that

$$E\,(y\zeta_\Lambda\,\alpha_{\Delta\setminus\Lambda}|\emptyset) - E\,(x\zeta_\Lambda\,\alpha_{\Delta\setminus\Lambda}|\emptyset) = E(y|\zeta_\Lambda\,\alpha_{\Delta\setminus\Lambda}) - E(x|\zeta_\Lambda\,\alpha_{\Delta\setminus\Lambda})$$

$$\leq E_+(y|\zeta_\Lambda\,\alpha_{\Delta\setminus\Lambda}) + B$$

whenever $\zeta_\Lambda\,\alpha_{\Delta\setminus\Lambda} \in \Omega_{r(\Phi)}$ and $\zeta(\Lambda) + \alpha(\Delta\setminus\Lambda) = N-1$. Here $E_+$ denotes the energy defined using $\Phi_+$ instead of $\Phi$, and $B$ is either chosen as in (6.11) or $-(N-1)$ times a lower bound of $\Phi$. Thus the expression on the r.h.s. of (6.31) has a lower bound of the form

$$\frac{1}{n+1}\, e^{-B} \limsup_{\Delta\uparrow S} \int m_\Delta(d\alpha) \int_{\Delta\setminus\Lambda} \sigma(dy) \exp[-E_+(y|\alpha)]$$

where the $m_\Delta$'s are certain probability measures on $\Omega$ with $m_\Delta(N(S)=N-1) = 1$.

This lower bound is infinite. In fact, since $\sigma$ is infinite it is sufficient to prove

$$\sup_{\alpha:\alpha(S)=N-1} \int \sigma(dy)\ (1-\exp[-E_{+}(y|\alpha)]) < \infty .$$

However, this follows from (6.12) because of the inequality

$$1 - \prod_{1}^{N-1} a_i \leq \sum_{1}^{N-1} (1-a_i) \quad (0 \leq a_i \leq 1) . \quad \lrcorner$$

Next let us verify condition (6.7) (V) under the assumption (C). This gives the first part of Theorem (6.14).

(6.32) Proposition: *Suppose condition* (C) *holds. Then*

$$z^{*}(\mu) \geq 1 \,/\, z(\mu)$$

*for all* $\mu \in \mathrm{ex}\ C \smallsetminus \{\varepsilon_{\emptyset}\}$ .

Proof: This result will follow from (6.29) and the definition of $z(\cdot)$ and $z^{*}(\cdot)$ once we have shown that there is a constant $K < \infty$ such that for all $\Delta \in \mathcal{S}$, $N \geq 1$, and $\omega \in \Omega_{r(\Phi)}$ with $Z_{\Delta,N}(\omega) > 0$

(6.33) $$N\ Z_{\Delta,N}(\omega) \,/\, Z_{\Delta,N-1}(\omega) \leq K + (N+1)\ Z_{\Delta,N+1}(\omega) \,/\, Z_{\Delta,N}(\omega) .$$

Applying Jensen's inequality to the r.h.s. of the identity in (6.28) we see that the l.h.s. of (6.33) has the upper bound

(6.34) $$N^{-1}\ Z_{\Delta,N}(\omega)^{-1} \int_{\{N(\Delta)=N\}} \pi(d\zeta) \int \zeta(dx)\, \exp[-E_{\Delta}(\zeta|\omega)] \int_{\Delta} \sigma(dy)\ \exp[-E(y|\zeta_{\Delta}\omega_{S\smallsetminus\Delta}\smallsetminus x)]$$

(cf. the proof of (5.34)). From condition (C) we obtain the estimate

$$\int_{\Delta} \sigma(dy)\ \exp[-E(y|\zeta_{\Delta}\ \omega_{S\setminus\Delta}\setminus x)] - \int_{\Delta} \sigma(dy)\ \exp[-E(y|\zeta_{\Delta}\ \omega_{S\setminus\Delta})]$$

$$= \int_{\Delta} \sigma(dy)\ \exp[-E(y|\zeta_{\Delta}\ \omega_{S\setminus\Delta}\setminus x)]\ (1-e^{-\Phi(x,y)}) \leq e^{B} \int \sigma(dy)(1-e^{-\Phi_{+}(x,y)}) = K < \infty.$$

In serting this into (6.34) we get the upper bound

$$N^{-1} \int \pi(d\zeta)\ \gamma_{\Delta,N}(\zeta|\omega)\ \zeta(\Delta)\ \{K + \int_{\Delta} \sigma(dy)\ \exp[-E(y|\zeta_{\Delta}\ \omega_{S\setminus\Delta})]\}$$

which, according to (6.28), is exactly the r.h.s. of (6.33). ┘

Now we assume that condition (D) is satisfied. Then there is an interger $R < \infty$ with

$$\Phi(x,y) = 0 \quad (|x-y| \geq R),$$

and it is natural to perform the limit $\Delta \uparrow S$ through a sequence $\Delta(n)$ for which $\Delta(n)$ and $S \setminus \Delta(n+1)$ have a distance apart not less than $R$. We choose

(6.35) $\qquad \Delta(n) = \{x \in S : |x| \leq nR\} \quad (n > 0)$

and assume that in the definition of $z(\cdot)$ and $z^*(\cdot)$ $\Delta$ runs through this sequence. By definition we have $\rho_+(\omega) < \infty$ for $\omega \in \Omega_0$, where

(6.36) $\qquad \rho_+(\omega) = \limsup_{n\to\infty} \omega(\Delta(n)) / \sigma(\Delta(n)).$

The next proposition shows that, under the hypothesis (D), $z^*(\mu) > 0$ for all $\mu \in \text{ex}\ C$ (recall that by definition these $\mu$ are supported on $\Omega_0$). Hence condition (D) implies (6.7) (V) and therefore the hypothesis of (6.10), thus proving the second part of (6.14).

(6.37) <u>Proposition:</u> *Assume* (D) *holds. Then for each* $\omega \in \Omega_0$ *there is a constant* $K(\omega) > 0$ *depending only on* $\rho_+(\omega)$ *and such that*

$$\limsup_{n\to\infty} Z_{\Lambda(n),\omega(\Lambda(n))+1}(\omega) \,/\, Z_{\Lambda(n),\omega(\Lambda(n))}(\omega) \geq K(\omega)$$

*for any sequence* $\Lambda(n)$ *in* $S$ *with* $\Lambda(n) \subset \Delta(n)$ *and for which* $\Delta(n) \setminus \Lambda(n) \subset W$ *for some* $W \in S$.

Proof: 1. First let us show

$$\liminf_{n\to\infty} \sigma(\Delta(n+1)) \,/\, \sigma(\Delta(n)) = 1 .$$

For otherwise we could find an $\varepsilon > 0$ and an integer $k \geq 1$ with $\sigma(\Delta(k)) > 0$ and

$$\sigma(\Delta(n+1)) \geq (1+\varepsilon)\,\sigma(\Delta(n)) \qquad (\, n \geq k\text{-}1 \,)$$

Let $a_{n+1} = \sigma(\Delta(n+1) \setminus \Delta(n))$. Then for all $n \geq k$ $a_{n+1} \geq \varepsilon \sum_k^n a_i$. By induction we obtain $a_{n+1} \geq \varepsilon(1+\varepsilon)^{n-k} a_k > 0$. This is impossible since $a_n \leq c\,\lambda(\Delta(n)) \leq c'\, n^d$ for some $c' < \infty$.

2. Because of (6.28) the proposition will be proved if we can find a constant $K(\omega) > 0$ depending only on $\rho_+(\omega)$ and such that for infinitely many $n$

$$\int_{\Lambda(n)} \sigma(dx) \exp[-E(x|\zeta)] \geq K(\omega)\;\rho_+(\omega)\;\sigma(\Lambda(n))$$

whenever $\zeta \in \Omega$ satisfies $\zeta(\Delta(n+1)) = \omega(\Delta(n+1))$. We assume $\rho_+(\omega) > 0$ (otherwise) the assertion is trivial) and consider the set $I$ of all integers $n$ with

$$\omega(\Lambda(n+1)) \leq 2\,\rho_+(\omega)\;\sigma(\Lambda(n)) .$$

By step 1. $I$ is infinite. Next we choose a constant $c$ with $\sigma \leq c\,\lambda$ and determine an $a > 0$ such that

$$2\,\rho_+(\omega)\;c\;v_a \leq 1/2$$

where $v_a = \lambda(x \in \mathbb{R}^d : |x| \leq a)$. We fix an $n \in I$ and let $\Lambda = \Lambda(n)$. Further, we fix a configuration $\zeta$ with $\zeta(\Delta(n+1)) = \omega(\Delta(n+1))$ and consider the set

$$V = \{x \in \Lambda : |x-y| > a \text{ for all } y \in \zeta\}.$$

Then

$$\begin{aligned}(6.38) \qquad \sigma(\Lambda \setminus V) &\leq \omega(\Delta(n+1))\ c\ v_a \\ &\leq 2\ \rho_+(\omega)\ \sigma(\Lambda)\ c\ v_a \\ &\leq \sigma(\Lambda)\ /\ 2\,.\end{aligned}$$

On the other hand we have for each $\alpha > 0$

$$\begin{aligned}&\alpha \quad \sigma(x \in V : E(x|\zeta) \geq \alpha) \\ &\leq \int_V \sigma(dx)\ E(x|\zeta) \vee 0 \\ &\leq \sum_{y \in \zeta_{\Delta(n+1)}} \int_{\{|x-y|>a\}} \sigma(dx)\ \Phi_+(x,y) \\ &\leq \omega(\Delta(n+1))\ C_a \leq 2\ \rho_+(\omega)\ C_a\ \sigma(\Lambda)\end{aligned}$$

and for the particular choice $\alpha = 8\ \rho_+(\omega)\ C_a$ we get

$$\sigma(x \in V : E(x|\zeta) \geq \alpha) \leq \sigma(\Lambda)\ /\ 4\,.$$

Combined with (6.38) this gives

$$\sigma(x \in \Lambda : E(x|\zeta) \leq \alpha) \geq \sigma(\Lambda)\ /\ 4$$

and thus

$$\int \sigma(dx) e^{-E(x|\zeta)} \geq e^{-\alpha}\ \sigma(\Lambda)\ /\ 4\,.$$

Hence $K(\omega) = e^{-\alpha} / 4\, \rho_+(\omega)$ has the desired property. ┘

Let us mention once more that the requirement $\mu(\Omega_{r(\Phi)}) = 1$ in the definition of $C$ is essential in the proof of Theorem (6.14) under the hypothesis (D), i.e., when $r(\Phi) = 0$. In the case $r(\Phi) > 0$ the same requirement was essential in the proof of Proposition (6.7), where it was needed for the implication (V) ⇒ (III), see (6.22) above. (Notice that the proof of Theorem (6.14) under hypothesis (C) is based on this implication). Therefore it is interesting to observe that this requirement is automatically satisfied in some particular situations. Let $C^*$ denote the class of all probability measures $\mu$ on $(\Omega, F)$ which satisfy the conditions of definition (1.68) except for the requirement $\mu(\Omega_{r(\Phi)}) = 1$. Clearly $\mu(\Omega_{(r(\Phi))}) = 1$ for all $\mu \in C^*$.

(6.39) Remark: *Suppose* $\Phi$ *is a nonnegative finite range potential with* $r(\Phi) = 0$. *Then* $\mu(\rho_+(.) < \infty) = 1$ *for all* $\mu \in C^*$ *and therefore* $C^* = C$.

Proof: Since (1.72) holds also for the class $C^*$ we may assume $\mu \in \text{ex}\, C^*$. Let $V \in S$ and choose a set $\Lambda \in S$ with $\Lambda \supset V$ and $\Phi(\alpha) = 0$ for $\alpha(V)\, \alpha(S \setminus \Lambda) > 0$. Assume $\mu(\rho_+(.) = \infty) > 0$. Then, using the same argument as in (6.31), we obtain from (1.74) (which clearly holds also when $C$ is replaced by $C^*$) that there exists an $\omega \in \Omega$ with $\rho_+(\omega) = \infty$ and with the following property: For $\sigma \otimes \pi$-a.a. $(x,\zeta) \in V \times \Omega$ with $u^\mu_\Lambda(\zeta) > 0$ the ratio $u^\mu_\Lambda(x\zeta) / u^\mu_\Lambda(\zeta)$ is the limit of the expressions

$$\frac{(n+1) \displaystyle\int_{\{N(\Delta \setminus \Lambda) = n\}} \pi(d\alpha)\, \exp[-E_\Delta(x\zeta_\Lambda\, \alpha_{\Delta \setminus \Lambda} | \omega)]}{\displaystyle\int_{\{N(\Delta \setminus \Lambda) = n\}} \pi(d\alpha) \int_{\Delta \setminus \Lambda} \sigma(dy)\, \exp[-E_\Delta(y\zeta_\Lambda\, \alpha_{\Delta \setminus \Lambda} | \omega)]}$$

when $\Delta \uparrow S$. Here $n = \omega(\Delta) - \zeta(\Lambda) - 1$. Using the estimate

$$E_\Delta(x\zeta_\Lambda\, \alpha_{\Delta \setminus \Lambda} | \omega) - E_\Delta(y\zeta_\Lambda\, \alpha_{\Delta \setminus \Lambda} | \omega) \le E(x | \zeta_\Lambda)$$

we see that this limit is infinite. Hence $u^\mu_\Lambda = 0$ $\pi$-a.s., contradicting the fact that $u^{\mu}_{\Lambda}$ is a probability density. ┘

If $\Phi$ is not necessarily nonnegative but shift-invariant and $\sigma = \lambda$ then the d-dimensional ergodic theorem asserts that $\rho_+(\cdot) < \infty$ $\mu$-a.s. for all shift-invariant $\mu$. Hence in this case we have at least $C^*_\theta = C_\theta$.

Next we consider the case $r(\Phi) > 0$. For geometrical reasons we have to confine ourselves to a one-dimensional position space.

(6.40) Remark: *Suppose* S *is an unbounded interval and there are constants* $0 < c_1 \le c_2 < \infty$ *with* $c_1 \lambda \le \sigma \le c_2 \lambda$ *on* S. *We assume that* $\Phi$ *is a potential with a hard care* $r = r(\Phi) > 0$ *such that*

$$B = \sup \{ |E(x|\omega)| : x\omega \in \Omega_{(r)} \} < \infty .$$

*For each configuration* $\omega \in \Omega_{(r)}$ *of hard rods of length* r *we let*

$$\ell(\omega) = \lambda(x \in S : |x-y| > r/2 \text{ for all } y \in \omega)$$

*denote the length of the region which is not covered by* $\omega$. *Then*

$$\mu(\ell(\cdot) = \infty) = 1$$

*for all* $\mu \in C^*$. *In particular,* $C^* = C$.

Proof: Let $q = 1 - (c_1/c_2)\, e^{-2B} < 1$ and $\mu \in \text{ex}\, C^*$. We show that the assumption

$$\mu(\ell(\cdot) < \infty) > 0$$

leads to the absurd conclusion

$$\mu(0 < \ell(\cdot) \le q\, \ell(\cdot) < \infty) = 1 .$$

First observe that $\ell(\cdot)$ is measurable with respect to $E_\infty$. Hence $\ell(\cdot) = \ell$ $\mu$-a.s.

for a suitable constant $0 < \ell < \infty$. Next we fix an interval $\Lambda = [a,b]$ in $S$. For any $\omega \in \Omega_{(r)}$ let

$$\ell^+_\Lambda(\omega) = \lambda(a-r/2 \leq x \leq b+r/2 : |x-y| > r/2 \quad \text{for all} \quad y \in \omega)$$

denote t free length for the hard rods in $\omega_\Lambda$ when $\omega_{S\setminus\Lambda}$ is fixed; similarly, let

$$\ell^-_\Lambda(\omega) = \lambda(a+r/2 \leq x \leq b-r/2 : |x-y| > r/2 \quad \text{for all} \quad y \in \omega_\Lambda$$

denote the length of the region which remains uncovered when $\omega_\Lambda$ is fixed. Clearly $\ell_\Lambda(\cdot) \uparrow \ell(\cdot)$ when $\Lambda \uparrow S$. In particular the claimed contradiction will follow once we have proved

$$u^\mu_\Lambda(\cdot) = 0 \quad \pi\text{-a.s.} \quad \text{on} \quad \{\ell^-_\Lambda(\cdot) > q\,\ell\} .$$

Because of (1.74) there is an $\omega \in \Omega_{(r)}$ with $\ell(\omega) = \ell$ and

$$u^\mu_\Lambda(\zeta) = \lim_{\Delta \uparrow S} \int \pi(d\alpha)\gamma_{\Delta,\omega(\Delta)}\ (\zeta_\Lambda\ \alpha_{\Delta\setminus\Lambda}|\omega)$$

for $\pi$-a.a. $\zeta$. We calculate this limit when $\ell^-_\Lambda(\zeta) > q\,\ell$. Writing $n = \omega(\Delta)$ we have

$$\int_{\{\zeta(\Lambda)+N(\Delta\setminus\Lambda) = n\}} \pi(d\alpha)\ \exp[-E_\Lambda(\zeta_\Lambda\ \alpha_{\Delta\setminus\Lambda}|\omega)]$$

$$\leq e^{nB}\ e^{-\sigma\ \Delta\setminus\Lambda)}\ [c_2(\ell^+_\Delta(\omega) - \ell^-_\Lambda(\zeta))_+]^{n-\zeta(\Lambda)}\ /\ (n - \zeta(\Lambda))!\quad .$$

Similarly,

$$\int_{\{N(\Delta) = n\}} \pi(d\alpha)\ \exp[-E_\Delta(\alpha_\Delta|\omega)] \quad \geq\ e^{-nB}\ e^{-\sigma(\Delta)}\ [c_1\ \ell^+_\Delta(\omega)]^n\ /\ n!\quad .$$

Thus $u^{\mu}_{\Lambda}(\zeta)$ is dominated by a multiple of

$$\lim_{n\to\infty} (1-q)^{-n} \, (1-\ell^{-}_{\Lambda}(\zeta) \,/\, \ell)^{n}_{+} \; n^{\zeta(\Lambda)}$$

which equals zero when $\ell^{-}_{\Lambda}(\zeta) \,/\, \ell > q$ . ∎

*Bibliographical notes:* The first attempt to find sufficient conditions for (6.1) can be found in Georgii (1976 b). Here we obtain far more general results by using different techniques. In the proof of (6.32) and (6.37) we use estimates which are due to R. L. Dobrushin and R. A. Minlos (1967) and J. Ginibre (see Ruelle (1969), p. 58). A somewhat different approach which leads to similar results was independently found by S. Goldstein (1976), see also the final chapter of M. Aizenman et al. (1978). In the latter reference , as well as in C. Preston (1979), the more general microcanonical case is investigated.

## § 7 Some further results on homogeneous models

In this section we consider again spatially homogeneous models. We confine ourselves to the discrete case and give only some hints concerning the continuous one. We will show that the activity function can be chosen to be shift-invariant, and then investigate its relation to the particle density. In section 7.2 we will discuss the connections between our basic identity

$$\text{(7.1)} \qquad \text{ex}\, C = \bigcup_z \text{ex}\, G(z)$$

and the classical problem in statistical physics of whether the canonical and the grand canonical ensemble are, in some sense, equivalent.

### 7.1 Properties of the activity function

Let $S = \mathbb{Z}^d$ be the d-dimensional integer lattice and $\Phi$ a shift-invariant potential. Condition (A) of § 5 is then satisfied. Consequently, (7.1) holds and the activity function $z(\cdot)$ is given by the limit in (5.14) (b).

(7.2) Remark: *For all* $x \in S$ *and* $\mu \in C$ *we have*

$$z(\theta_x(.)) = z(.) \quad \mu\text{-a.s.}$$

*Thus there is a shift-invariant function* $z_\theta(.)$ *with* $z_\theta(.) = z(.)$ $\mu$-a.s. *for all* $\mu \in C$.

Proof: Let $x \in S$. Then for all $\Lambda, \omega$, a, b we have

$$Z_{\Lambda,N((\theta_x\omega)_\Lambda)+1_b-1_a}(\theta_x\omega) \,/\, Z_{\Lambda,N((\theta_x\omega)_\Lambda)}(\theta_x\omega)$$

$$= Z_{\Lambda-x,N(\omega_{\Lambda-x})+1_b-1_a}(\omega) \,/\, Z_{\Lambda-x,N(\omega_{\Lambda-x})}(\omega)$$

If $\Lambda \uparrow S$ then also $\Lambda - x \uparrow S$, and (5.14) (b) holds for both sequences. This proves the first assertion. $z_\theta(\cdot)$ then can be defined by

$$z_\theta(a,\omega) = \liminf_{\Delta \uparrow S} |\Delta|^{-1} \sum_{x \in \Delta} z(a,\theta_x \omega),$$

where $\Delta$ runs through a sequence of cubes. ┘

From the above remark we obtain a second proof of (3.27):

(7.3) Corollary: $\mathrm{ex}\, C_\theta = \bigcup_{z \in A} \mathrm{ex}\, G_\theta(z)$, *and* (3.27) (ii) *and* (iii) *hold also when* $z(\rho(\cdot))$ *is replaced by* $z_\theta(\cdot)$. *Thus if* $\Phi$ *satisfies* (3.1) *then*

$$z(\cdot) = z_\theta(\cdot) = z(\rho(\cdot)) \quad \mu\text{-a.s.}$$

*for all* $\mu \in C_\theta$.

Proof: Let $\mu \in \mathrm{ex}\, C_\theta$. Then $\mu$ is ergodic. Hence $z_\theta(\cdot)$ and $z(\cdot)$ are a.s. constant. Thus (5.14) (d) implies $\mu \in \bigcup_z G(z)$. Conversely, any $\mu \in \bigcup_z \mathrm{ex}\, G(z)$ is ergodic and therefore extreme in $C_\theta$. (3.27) (ii) and (iii) follow from (5.14) (d) and (e) by observing that when $\mu$ is shift-invariant then $\mu_z = \mu(\cdot | z_\theta(\cdot) = z)$ is also shift-invariant for a.a. $z$. ┘

It is obvious that (7.2) and the first part of (7.3) hold also in the continuous model whenever $\Phi$ is a shift-invariant potential for which (7.1) is true. Moreover, if $\Phi$ is a finite range pair potential having a hard core and bounded from below (thus condition (C) is satisfied with $\sigma = \lambda$) then it can be verified that $z(\omega)$ is a.s. a function of the particle density $\rho(\omega)$: The continuous counterpart of the function $Q(\cdot)$ (defined in (3.10)) exists (see, for example, Lanford (1973)), and Dobrushin and Minlos (1967) have shown that this function is differentiable, its derivative at $\rho(\omega)$ being the negative of the logarithm of the limit in (6.10) (b). (Actually, both these papers do not include boundary conditions, but in the hard core case it is not difficult to do so). Here we will state and prove the discrete analogue of Theorem 2 in Dobrushin/Minlos (1967); this result shows that for many shift-invariant potentials the assertion (5.14) (b) remains true when $N(\omega_\Lambda)$ is replaced by any $L_\Lambda \in A_\Lambda$ such that $|L_\Lambda - N(\omega_\Lambda)| / |\Lambda| \to 0$ and where the boundary condition $\omega_{S\setminus\Lambda}$ is replaced by an arbitrary $\alpha_{S\setminus\Lambda}$. However, we will confine ourselves to the particular sequence $(\Lambda_n)$ of cubes defined by (3.8). We consider again the function $\rho \to z(\rho) = z(\cdot,\rho)$ introduced in (3.9) and discussed thereafter.

(7.4) Proposition: *Suppose* $\Phi$ *is a shift-invariant potential with*

$$\sum_{A\ni 0} (|A| - 1) \, \| \Phi(A,\cdot) \| < \infty .$$

*Let* $a,b \in F$ *and* $\rho \in A_1$ *be such that* $\rho(b) > 0$. *Then*

$$z(a,\rho) / z(b,\rho) = \lim_{n\to\infty} Z_{\Lambda_n, L_n + 1_b - 1_a}(\alpha_n) / Z_{\Lambda_n, L_n}(\alpha_n)$$

*for any sequence* $(L_n)$ *with* $L_n \in A_{\Lambda_n}$ *and* $L_n / |\Lambda_n| \to \rho$ *and any sequence* $(\alpha_n)$ *in* $\Omega$.

Assuming the hypotheses of (7.4) we thus obtain from (5.14) (b): For all $\mu \in C$ and $\mu$-a.a. $\omega$ the equation

$$z(\omega) = z(\rho) \tag{7.5}$$

holds, where $\rho$ is an arbitrary limit point of the sequence $N(\omega_{\Lambda_n}) / (\Lambda_n)$, $n \geq 1$.

Proof: It is easily seen from (5.28) that the assertion is valid when $\rho(a) = 0$. Thus we assume $\rho(a) > 0$. We fix a sufficiently large cube $\Lambda = \Lambda_n$, let $L = L_n$ and introduce the abbreviations

$$N = L(b), \quad L + j = L + j\, 1_b - j\, 1_a ,$$

and

$$Z_{L+j} = Z_{\Lambda, L+j}(\alpha) ,$$

$j \in \mathbb{Z}$. The assumptions on $\Phi$ imply that (5.37) holds with a certain constant $M$.

Thus we have

$$(N+j)Z_{L+j} / Z_{\Lambda+j-1} \leq M + (N+j+1)Z_{L+j+1} / Z_{L+j}$$

and therefore (by iteration) for $k \geq 0$

$$(N-k+1)Z_{L-k+1} / Z_{L-k} \leq (N+1)Z_{L+1} / Z_L \quad + \quad k\,M$$

Hence

$$Z_{L+1} / Z_{L-k} \leq \prod_{j=0}^{k} [(N+1)\, Z_{L+1} / Z_L + j\,M] / (N-j+1)$$

$$\leq (\frac{N+1}{N-k+1})^{k+1} [Z_{L+1} / Z_L + \frac{k}{N+1} M]^{k+1} .$$

This inequality can be rewritten as

$$Z_{L+1} / Z_L + \frac{k}{N+1} M$$

$$\geq \frac{N-k+1}{N+1} \exp[\frac{|\Lambda|}{k+1} \{|\Lambda|^{-1} \log Z_{L+1} - |\Lambda|^{-1} \log Z_{L-k}\}] .$$

In the limit $L / |\Lambda| \to \rho$, $k / |\Lambda| \to \delta > 0$ (3.10) gives

$$\liminf\ Z_{L+1} / Z_L + \frac{\delta}{\rho(b)} M \geq \frac{\rho(b)-\delta}{\rho(b)} \exp[-\{Q(\rho+\delta 1_a-\delta 1_b) - Q(\rho)\} / \delta]$$

Letting $\delta \to 0$ we finally get from (3.14)

$$\liminf Z_{L+1} / Z_L \geq z(a,\rho) / z(b,\rho) .$$

Now the proposition follows by interchanging the rôles of $a$ and $b$. ┘

## 7.2 The equivalence of ensembles

We continue to consider the spatially homogeneous discrete model. Let us assume that the (shift-invariant) potential $\Phi$ satisfies the condition

$$(7.6) \qquad \sum_{A \ni 0} (|A|-1) \; \| \Phi(A,\cdot) \| < \infty$$

of proposition (7.4); thus (7.5) holds, i.e., the activity function depends only on the particle densities.

In statistical physics the family of all probability spaces of the form $(\Omega_{\Lambda,L}, \gamma_{\Lambda,L}(\cdot|\omega))$ is called the canonical ensemble; similarly, the probability spaces $(\Omega_\Lambda, \gamma_\Lambda{}^z(\cdot|\omega))$ form the grand canonical ensemble for $z$. J. W. Gibbs (1902) (who introduced these concepts) already believed that these ensembles are, in some sense, asymptotically equivalent when $|\Lambda| \to \infty$ and simultaneously $L / |\Lambda| \to \rho$. (Originally boundary conditions $\omega$ were not considered). Such an asymptotic equivalence of course presupposes a unique correspondence of the parameters $\rho$ and $z$. This correspondence is based on equation (3.12), the latter being the first form in which an equivalence was rigorously established. Therefore we introduce the set

$$(7.7) \qquad A_1^* = \{\rho \in A_1 : \{\rho\} = \{\rho' \in A_1 : z(\rho') = z(\rho)\}\}$$

of all those $\rho$ in a neighbourhood of which the specific free Helmholtz energy $Q(.)$ (being defined in (3.9)) is strictly concave. For the sake of convenience we only consider the particular sequence $(\Lambda_n)$ introduced in (3.8) whenever we need to take the limit $\Lambda \uparrow S$.

(7.8) Remark: *Let* $\rho \in A_1$. *Then* $\rho \in A_1^*$ *if and only if*

$$\rho = \lim_{n\to\infty} N(X_{\Lambda_n}) \,/\, |\Lambda_n| \quad \mu\text{-a.s.}$$

*for all* $\mu \in G(z(\rho))$.

We will prove this in (7.17).

There are at least three possible formulations for an asymptotic equivalence of ensembles corresponding to a given particle density vector $\rho \in A_1^*$; let us discuss these possibilities now.

1. The most classical way of expressing an equivalence of ensembles for $\rho$ is the following condition $(E_\rho)$:

$(E_\rho)$ For each sequence $(L_n)$ with $L_n \in A_{\Lambda_n}$ and $L_n / |\Lambda_n| \to \rho$, for any sequence $(\omega_n)$ in $\Omega$, and arbitrary $V \in S$ and $\zeta \in \Omega_V$.

(7.9) $$\lim_{n\to\infty} | \gamma_{\Lambda_n, L_n} (X_V = \zeta | \omega_n) - \gamma_{\Lambda_n}^{z(\rho)} (X_V = \zeta | \omega_n) | = 0 .$$

We will show below that $(E_\rho)$ is satisfied whenever $|G(z(\rho))| = 1$. But first let us observe that (7.9), for "typical" sequences $(L_n)$ and $(\omega_n)$, follows immediately from (7.1).

(7.10) Remark: *Suppose* $\rho \in A_1$ *and* $\mu \in G(z(\rho))$. *Then*

$$\lim_{\Lambda\uparrow S} |\gamma_{\Lambda, N(\omega_\Lambda)} (X_V = \zeta|\omega) - \gamma_\Lambda^{z(\rho)} (X_V = \zeta|\omega) | = 0$$

*for* $\mu$*-a.a.* $\omega$ *and arbitrary* $V \in S$ *and* $\zeta \in \Omega_V$.

Proof: We may assume $\mu \in \text{ex}\, G(z(\rho))$. Then $\mu \in \text{ex}\, C$, and the assertion follows from (1.32) (b) and (1.34). ┘

2. The next fomulation of equivalence (which is closely connected to the first) is in terms of the sets $C_\infty(\rho)$ and $G_\infty(z)$: We let $C_\infty(\rho)$ denote the (weakly) closed convex hull of the set of all probability measures on $(\Omega, F)$ which are a weak limit point of a sequence of the form $(\gamma_{\Lambda_n, L_n}(\cdot|\omega_n))_{n\geq 1}$ with $\omega_n \in \Omega$ and $L_n \in A_{\Lambda_n}$ such that $L_n / |\Lambda_n| \to \rho$. (Here we think of $\gamma_{\Lambda, L}(\cdot|\omega)$ as a measure on $\Omega$ which is concentrated on $\{X_{S\setminus\Lambda} = \omega_{S\setminus\Lambda}\}$). Similarly, $G_\infty(z)$ is defined as the closed convex hull of all possible limit points of sequences of the form $(\gamma_{\Lambda_n}^z(\cdot|\omega_n))_{n\geq 1}$.

We then may express the equivalence of ensembles for a given $\rho \in A_1^*$ by the identity

(7.11) $$C_\infty(\rho) = G_\infty(z(\rho)) .$$

Obviously $(E_\rho)$ implies (7.11). A combination of (7.4) with the ideas used in (5.17) gives

(7.12) <u>Proposition:</u> *For each* $\rho \in A_1$ *we have*

$$C_\infty(\rho) \subset G_\infty(z(\rho)) .$$

*In particular,* (7.11) *holds whenever* $\rho \in A_1^*$.

This will be proved in (7.19).

(7.13) <u>Corollary:</u> *Let* $\rho \in A_1$ *and suppose that* $|G(z(\rho))| = 1$. *Then condition* $(E_\rho)$ *holds.*

<u>Proof:</u> We choose $(L_n)$, $(\omega_n)$, $V,\zeta$ as in $(E_\rho)$. In order to prove (7.9) it is sufficient to realize that every infinite subset $J$ of the integers contains an infinite subset $I$ such that (7.9) holds when $n$ tends to infinity through the set $I$. If such a $J$ is given then the diagonal method gives us an infinite set $I$ such that for all $\Delta \in S$ and $\alpha \in \Omega_\Delta$ the limits

$$\lim_{I \ni n \to \infty} \gamma_{\Lambda_n, L_n} (X_\Delta = \alpha | \omega_n)$$

and

$$\lim_{I \ni n \to \infty} \gamma_{\Lambda_n}^{z(\rho)} (X_\Delta = \alpha | \omega_n)$$

exist. These limits define two probability measures $\mu$ and $\nu$ which by definition belong to $C_\infty(\rho)$ and $G_\infty(z(\rho))$ respectively. Now (7.12) asserts that both measures belong to $G_\infty(z(\rho))$ which, as we will show below, coincides with $G(z(\rho))$. Hence $\mu = \nu$ and (7.9) holds when $n$ is restricted to $I$. ┘

Let us mention that $|G(z(\rho))| = 1$ only when $\rho \in A_1^*$. This follows from

(7.8) by approximating an arbitrary extreme point $\rho'$ of $\{z(\cdot) = z(\rho)\}$ by a sequence $(\rho_n)$ in $A_1^*$, amd then observing that any limit point $\mu$ of a sequence $(\mu_n)$ with $\mu_n \in G_\theta(z(\rho_n))$ belongs to $G_\theta(z(\rho))$ and satisfies $\int N(\cdot,\omega_0)\mu(d\omega) = \rho'$.

3. A third formulation of equivalence is obtained from (7.11) by noting that the sets $C_\infty(\rho)$ and $G_\infty(z)$ are closely related to $C$ and $G(z)$. Let $C_\infty$ denote the closed convex hull of the union of all sets $C_\infty(\rho)$, $\rho \in A_1$.

(7.14) Remark: $C_\infty = C$ *and* $G_\infty(z) = G(z)$ *for all* $z \in A$.

This remark holds in general, without any homogeneity assumption on $S$ or $\Phi$, and will be proved in (7.18). Combined with (7.11) it suggests considering the set

$$C(\rho) = \{\mu \in C : \rho = \lim_{n\to\infty} N(X_{\Lambda_n}) / |\Lambda_n| \ \mu\text{-a.s.}\}$$

instead of $C_\infty(\rho)$. Because of (1.32) (b) and the easily verified identity

$$\text{ex } C(\rho) = C(\rho) \cap \text{ex } C$$

we see immediately that

(7.15) $$C(\rho) \subset C_\infty(\rho).$$

However, in general $C(\rho) \neq C_\infty(\rho)$. Indeed, a compactness argument gives $C_\infty(\rho) \neq \emptyset$, but an example of a $\Phi$ and $\rho$ for which $C(\rho) = \emptyset$ exists (see Higuchi (1979), for instance). Nevertheless, for $\rho \in A_1^*$ we get from (7.8), (7.15), (7.12), and (7.14) the equation $C(\rho) = C_\infty(\rho)$ and therefore the following version of the equivalence of ensembles:

(7.16) Remark: *If* $\rho \in A_1^*$ *then* $C(\rho) = G(z(\rho))$.

Let us notice that (7.16) is also a direct consequence of (7.1), or more precisely of statement (5.14) (d) combined with (7.5). Moreover, if we redefine $C(\rho)$ for $\rho \notin A_1^*$ in an appropriate way then under the hypothesis (7.5) the assertion

$$C(\rho) = G(z(\rho)) \quad \text{for all} \quad \rho \in A_1$$

is equivalent to (7.1). (This shows in which sense (7.1) expresses an equivalence of ensembles). It is not difficult to verify that a suitable definition of $C(\rho)$ is obtained by letting $C(\rho)$ denote the set of all $\mu \in C$ for which with probability 1 every limit point of the sequence $(N(X_{\Lambda_n}) / |\Lambda_n|)$ belongs to the set $\{\rho' : z(\rho') = z(\rho)\}$. (This coincides with our previous definition when $\rho \in A_1^*$).

We now give the proofs left outstanding.

(7.17) Proof of Remark (7.8): Let $\rho \in A_1^*$ and $\mu \in G(z(\rho))$. Then (5.14) (d) together with (7.5) shows that with probability 1 $\rho$ is the unique limit point of the sequence $(N(X_{\Lambda_n}) / |\Lambda_n|)$; thus this sequence converges a.s. to $\rho$. (A more direct proof of this conclusion can be based on (3.4) and (3.10); cf. the lemma in V § 1 of Minlos (1968)). Conversely, using the argument sketched after the proof of (7.13) we see that $\text{ex}\ G_\theta(z(\rho))$ contains measures with different particle density distributions whenever $\rho \notin A_1^*$. ┘

(7.18) Proof of Remark (7.14): (1.32) (b) shows that $\text{ex}\ C \subset C_\infty$; thus the Krein-Milman theorem gives $C \subset C_\infty$. Conversely, suppose that $\mu$ is the weak limit of a sequence $\mu_n = \gamma_{\Lambda_n, L_n}(\cdot|\omega_n)$, $n \geq 1$. For any $V \in S$ and $\zeta \in \Omega_V$ and sufficiently large $n$ the consistency property (1.26) implies

$$\mu_n(X_V = \zeta) = \int_{\{N(X_V) = N(\zeta)\}} \gamma_{V,N(\zeta)}(\zeta|\cdot)\, d\mu_n .$$

Passing to the limit $n \to \infty$ and noting that because of (1.12) (ii) $\gamma_V(\zeta|\cdot)$ is continuous we get $\mu \in C$. The assertion $G_\infty(z) = G(z)$ is proved similarly. ┘

(7.19) Proof of Proposition (7.12): 1. First we need to show that Proposition (7.4) remains valid when the sequence $(\Lambda_n)$ is replaced by a sequence of the form $(\Lambda_n \setminus V)$ for some $V \in S$. A glance at the proof of (7.4) shows that this more general statement is also true, provided we can prove that (3.10) continuous to hold under this substitution. Let us verify this extension of (3.10). We fix a set $\Lambda = \Lambda_n \supset V$ and $L \in A_{\Lambda \setminus V}$, $M \in A_V$, and $\alpha \in \Omega$. Repeating the argument in step 2 of (3.33) we get the estimate

$$Z_{V,M}(\alpha)\ Z_{\Lambda \setminus V,L}(\alpha) \le Z_{\Lambda,L+M}(\alpha)\ e^{2r(V)+2r(\Lambda \setminus V)}$$

where $r(\cdot)$ is defined by (3.32). In the limit $n \to \infty$ we have $r(\Lambda \setminus V) / |\Lambda \setminus V| \to 0$ ; thus (3.10) gives

$$\limsup |\Lambda \setminus V|^{-1} \log Z_{\Lambda \setminus V,L}(\alpha) \le Q(\rho)$$

whenever $L / |\Lambda \setminus V| \to \rho$. In order to obtain the opposite inequality for the lim inf we fix an integer $k$ and write $\Delta = \Lambda_k$. For sufficiently large $\Lambda = \Lambda_n$ we put as many disjoint translates of $\Delta$ as possible into $\Lambda \setminus V$, say $m$ of these, and choose the largest $N \in A_\Delta$ for which $m\,N \le L$. Then again the argument in step of (3.33) gives

$$Z_{\Lambda \setminus V,L}(\alpha) \ge (Z_{\Delta,N}(\alpha) e^{-2r(\Delta)})^m\ e^{-(|\Lambda \setminus V| - m\,|\Delta|)\,\|\Phi\|}$$

Letting $n \to \infty$ and $L / |\Lambda \setminus V| \to \rho$ we obtain

$$\liminf |\Lambda \setminus V|^{-1} \log Z_{\Lambda \setminus V,L}(\alpha) \ge$$

$$\ge \min_{|N - [\rho|\Delta|]| \le 1} |\Delta|^{-1} \log Z_{\Delta,N}(\alpha) - 2r(\Delta) / |\Delta| .$$

The conclusion that we wanted now follows by taking the limit $k \to \infty$. (Recall that the convergence in (3.10) is uniform in $\rho$).

2. Suppose now $\mu \in C_\infty(\rho)$, say

$$\mu = \lim_{I \ni n \to \infty} \gamma_{\Lambda_n,L_n}(\cdot|\alpha_n) .$$

An examination of the proof of (5.17) readily shows that for all $\Delta \in S$, $\zeta \in \Omega_\Delta$, and $\mu$-a.a. $\omega$

$$\mu_\Delta(\zeta|\omega) \,/\, \mu_\Delta(\omega_\Delta|\omega) = \lim_{V\uparrow S} \mu(X_\Delta = \zeta, X_{V\setminus\Delta} = \omega_{V\setminus\Delta}) \,/\, \mu(X_V = \omega_V)$$

$$= \exp[E_\Delta(\omega_\Delta|\omega) - E_\Delta(\zeta|\omega)] \lim_{V\uparrow S} \limsup_{I\ni n\to\infty}$$

$$Z_{\Lambda_n\setminus V, L_n - N(\zeta\omega_{V\setminus\Delta})}(\omega_V(\alpha_n)_{S\setminus V}) \,/\, Z_{\Lambda_n\setminus V, L_n - N(\omega_V)}(\omega_V(\alpha_n)_{S\setminus V})$$

From 1. above we get that the double limit is exactly

$$\prod_{a\in F} z(a,\zeta)^{N(a,\zeta) - N(a,\omega_\Delta)} .$$

This gives $\mu_\Delta(\zeta|\omega) = \gamma_\Delta^{z(\rho)}(\zeta|\omega)$ and thus $\mu \in G(z(\rho)) = G_\infty(z(\rho))$ .

3. If $\rho \in A_1^*$ then (7.14), (7.8) and (7.15) lead to the inclusion $G_\infty(z(\rho)) \subset C_\infty(\rho)$ . ┘

Let us conclude with some remarks on continuous models. Suppose $\sigma = \lambda$ and $\Phi$ is a shift-invariant pair potential with a hard core satisfying (6.12). In this situation the continuous analogue of (7.4) was proved by Dobrushin/Minlos (1967). An inspection of their proof shows that any sequence $(\Lambda_n)$ is allowed for which the continuous counterpart of (3.10) is valid. It follows from Lanford (1973), e.g., that only mild regularity properties of $(\Lambda_n)$ are required for this to hold; in particular, we can choose sequences of the form $\Lambda_n = \{|\cdot| \le n\} \setminus V$, $n \ge 1$ , $V \in S$. This, together with the estimates in the proof of (6.19), shows that Proposition (7.12) has a continuous analogue. Moreover, using a compactness argument similar to that in section 4 of Dobrushin (1969) we may conclude that for continuous hard core models the counterpart of $(E_\rho)$ is also true whenever $|G(z(\rho))| = 1$ . Further, if $\Phi$ has no hard core but satisfies (6.13) then (7.12) holds at least in the case when only the empty boundary condition $\omega_n = \emptyset$ is allowed in the definition of $C_\infty(\rho)$ . This can be obtained using the same arguments as in the hard core case.

*Bibliographical notes:* Under conditions which are stronger than those of (7.13) the equivalence of ensembles in the sense of $(E_\rho)$ was proved by R. L. Dobrushin and B. Tirozzi (1977). They derived it from a local central limit theorem for the particle numbers; this idea which is due to A. Khinchin has also been used in continuous hard core models, see A. M. Halfina (1969) and R. A. Minlos and A. Khaitov (1972, 1975). In the discrete case a concept of equivalence related to $(E_\rho)$ was studied by A. Martin-Löf (1977).

## References


Aizenman, M., Gallavotti, G., Goldstein, S., Lebowitz, J.L.

(1976) Stability and Equilibrium States of Infinite Classical Systems. Commun. math. Phys. 48, 1-14

Aizenman, M., Goldstein, S., Gruber, C., Lebowitz, J.L., Martin, P.

(1977) On the Equivalence between KMS-States and Equilibrium States for Classical Systems. Commun. math Phys. 53, 2o9-22o

Aizenman, M., Goldstein, S., Lebowitz, J.L.

(1978) Conditional Equilibrium and the Equivalence of Microcanonical and Grandcanonical Ensembles in the Thermodynamic Limit. Commun. math. Phys. 62, 279-3o2

Aldous, D., Pitman, J.

(1977) On the Zero-One Law for Exchangeable Events. (to appear)

Dobrushin, R.L.

(1968) Description of a Random Field by means of its Conditional Probabilities and Conditions of its Regularity. Theory Probability Appl. 13, 197-224

(1968) The Problem of Uniqueness of a Gibbsian Random Field and the Problem of Phase Transitions. Functional Anal. Appl. 2, 3o2-312

(1969) Gibbsian Random Fields. The General Case. Functional Anal. Appl. 3, 22-28

(197o) Gibbsian Random Fields for Particles without Hard Core. Theor. math. Physics 4, 7o5-719

Dobrushin, R.L., Minlos, R.A.

(1967) Existence and Continuity of Pressure in Classical Statistical Physics. Theory Probability Appl. 12, 535-559

Dobrushin, R.L., Tirozzi, B.

(1977) The Central Limit Theorem and the Problem of Equivalence of Ensembles. Commun. math. Phys. 54, 173-192

Dubins, L.E., Freedman, D.A.

(1979) Exchangeable Processes Need Not Be Mixtures of Independent, Identically Distributed Random Variables. Z. Wahrscheinlichkeitstheorie u. verw. Gebiete 48, 115-132

Dunford, N., Schwartz, J.T.

(1958) Linear Operators, Vol. I, New York: Interscience 1958

Dynkin, E.B.

(1978) Sufficient Statistics and Extreme Points. Ann. Prob. 6, 7o5-73o

Feller, W.

(1966) An Introduction to Probability Theory and Its Applications, Vol. II. New York: John Wiley 1966

de Finetti, B.

(1931) Funzione caratteristica di un fenomeno aleatorio. Atti della R. Accademia Nazionale dei Lincei, Ser. 6, Memorie, Classe di Scienze Fisiche, Matematiche e Naturali vol. 4, 251-3oo

Föllmer, H.

(1973) On Entropy and Information Gain in Random Fields. Z. Wahrscheinlichkeitstheorie u. verw. Gebiete 26, 2o7-217

(1975) Phase Transition and Martin Boundary. Seminaire de Probabilitès Strasbourg IX, Springer Lecture Notes in Mathematics 465

Freedman, D.A.

(1962) Invariants under Mixing which Generalize de Finetti's Theorem. Ann. Math. Stat. 33, 916-923

(1963) Invariants under Mixing which Generalize de Finetti's Theorem: Continuous Time Parameter. Ann. Math. Stat. 34, 1194-1216

Fritz, J.

(1978) Stochastic Dynamics of Two-Dimensional Infinite Particle Systems. Preprint of the Math. Institute of the Hung. Acad. of Sciences No. 9/1978 (to appear in J.Statist.Phys.)

Fritz, J., Dobrushin, R.L.

(1977) Non-Equilibrium Dynamics of Two-Dimensional Infinite Particle Systems with a Singular Interaction. Commun. math. Phys. 57, 67-82. Erratum: Commun. math. Phys. 65, 96

Georgii, H.O.

(1975) Canonical Gibbs States, their Relation to Gibbs States, and Applications to Two-Valued Markov Chains. Z. Wahrscheinlichkeitstheorie u. verw. Gebiete 32, 277-3oo

(1976) On Canonical Gibbs States, Symmetric and Tail Events. Z. Wahrscheinlichkeitstheorie u. verw. Gebiete 33, 331-341

(1976) Canonical and Grand Canonical Gibbs States for Continuum Systems. Commun. math. Phys. 48, 31-51

Gibbs, J.W.

(19o2) Elementary Principles in Statistical Mechanics. Yale University Press 19o2

Glötzl, E.

(1977) On the Singularity of (Σ) Point Processes. (to appear in Math. Nachrichten)

(1978) Gibbsian Description of Point Processes. Proceedings of the Colloquium on Point Processes, Keszthely 1978, North Holland (forthcoming)

Goldstein, S.

(1976) Canonical Gibbs States. (Lectures at the Institute for Advanced Studies, the Hebrew University of Jerusalem)

Halfina, A.M.

(1969) The Limiting Equivalence of the Canonical and Grand Canonical Ensembles (Low Density Case). Math. USSR Sbornik 9, 1-52

Hewitt, E., Savage, L.J.

(1955) Symmetric Measures on Cartesian Products. Trans. Amer. Math. Soc. 8o, 47o-5o1

Higuchi, Y.

(1979) On the Absence of Non-Translationally Invariant Gibbs States for the Two-Dimensional Ising Model. (to appear)

Höglund, T.

(1974) Central Limit Theorems and Statistical Inference for Finite Markov Chains. Z. Wahrscheinlichkeitstheorie u. verw. Gebiete 29, 123-151

Holley, R.

(1971) Pressure and Helmholtz Free Energy in a Dynamic Model of a Lattice Gas. Proc. Sixth Berkeley Symp. Prob. math. Statist. 3, 565-578

Kakutani, S.

(1948) On Equivalence of Infinite Product Measures. Ann. of Math. II. Ser. 49, 214-224

Kallenberg, O.

(1978) On Conditional Intensities of Point Processes. Z. Wahrscheinlichkeitstheorie u. verw. Gebiete 41, 2o5-22o

Kerstan, J., Matthes, K., Mecke, J.

(1974) Unbegrenzt teilbare Punktprozesse. Akademie-Verlag, Berlin 1974

Kolmogorov, A.N.

(1937) Zur Umkehrbarkeit der statistischen Naturgesetze. Math. Ann. 113, 766-772

Kozlov, O.K.

(1976) Gibbsian Description of Point Random Fields. Theory Prob. Appl. 21, 339-355

Lanford, O.E.

(1973) Entropy and Equilibrium States in Classical Statistical Mechanics. In: Statistical Mechanics and Mathematical Problems, Battelle Seattle 1971 Rencontres (ed. A. Lenard) Springer Lecture Notes in Physics 2o

(1975) Time Evolution of Large Classical Systems. In: Dynamical Systems, Theory and Applications (ed. J. Moser), Springer Lecture Notes in Physics 38

Lanford, O.E., Ruelle, D.

(1969) Observables at Infinity and States with Short Range Correlations in Statistical Mechanics. Commun. math. Phys. 13, 194-215

Lang, R.

(1977) Unendlich-dimensionale Wienerprozesse mit Wechselwirkung. Z. Wahrscheinlichkeitstheorie u. verw. Gebiete 38, 55-72 & 39, 277-299, Addendum: see Shiga (1979)

Lauritzen, S.L.

(1974) Suffiency, Prediction and Extreme Models. Scand. J. Statist. 1, 128-134

Liggett, Th. M.

(1971) Existence Theorems for Infinite Particle Systems. Trans. Amer. Math. Soc. 165, 471-481

(1976) Coupling the Simple Exclusion Process. Annals Prob. 4, 339-356

(1977) The Stochastic Evolution of Infinite Systems of Interacting Particles. Ecole d'Eté de Probabilités de Saint-Flour VI-1976, Springer Lecture Notes in Mathematics 598

Logan, K.G.

(1974) Time Reversible Evolutions in Statistical Mechanics. Cornell University, Ph. D. thesis

Marchioro, C., Pellegrinotti, A., Pulvirenti, M.

(1978) Selfadjointness of the Liouville Operator for Infinite Classical Systems. Commun. math. Phys. 58, 113-129

Martin-Löf, A.

(1977) The Equivalence of Ensembles and Gibbs Phase Rule for Classical Lattice Systems (to appear in J. Statist. Phys. 1979)

Martin-Löf, P.

(1974) Repetitive Structures and the Relation between Canonical and Microcanonical Distributions in Statistics and Statistical Mechanics. Proceedings of a Conference on Foundational Questions in Statistical Inference, Memoirs No. 1, 271-293, Dept. Theoret. Statist., Inst. Math., Univ. Aarhus

Matthes, K., Warmuth, W., Mecke, J.

(1977) Bemerkungen zu einer Arbeit von Nguyen Xuan Xanh und Hans Zessin. (to appear in Math. Nachrichten)

Mecke, J.

(1976) A characterization of Mixed Poisson Processes. Rev. Roum. Math. Pures et Appl. 21, 1355-136o

Meyer, P.A.

(1966) Probabilité et Potentiel. Paris: Hermann 1966

Minlos, R.A.

(1968) Lectures on Statistical Physics. Russian Math. Surveys 23, 137-194

Minlos, R.A., Khaitov, A.

(1972) Equivalence in the Limit of Thermodynamic Ensembles in the Case of One-Dimensional Classical Systems. Funct. Anal. Appl. 6, 337-338

Minlos, R.A., Khaitov, A.

(1975) Limiting Equivalence of Thermodynamic Ensembles in Case of One-Dimensional Systems. Trans.Moscow Math. Soc. 32, 143-18o

Moulin Ollagnier, J., Pinchon, D.

(1977) Free Energy in Spin-flip Processes is Non-increasing. Commun. math. Phys. 55, 29-35

Nawrotzki, K.

(1962) Ein Grenzwertsatz für homogene zufällige Punktfolgen. Math. Nachrichten 24, 2o1-217

Neveu, J.

(1977) Processus Ponctuels. Ecole d'Eté de Probabilités de Saint-Flour VI-1976, Springer Lecture Notes in Mathematics 598

Nguyen Xuan Xanh, Zessin, H.

(1976) Integral and Differential Characterizations of the Gibbs Process. (to appear in Math. Nachrichten)

(1977) Martin-Dynkin Boundary of Mixed Poisson Processes. Z. Wahrscheinlichkeitstheorie u. verw. Gebiete 37, 191-2oo

Papangelou, F.

(1974) The Conditional Intensity of General Point Processes and an Application to Line Processes. Z. Wahrscheinlichkeitstheorie u. verw. Gebiete 28, 2o7-226

Pitman, J.

(1978) An Extension of de Finetti's Theorem. Proc. of the Seventh Conference on Stochastic Processes and Applications, Adv. Appl. Prob. 1o, 268-27o

Preston, C.

(1974) Gibbs States on Countable Sets. Cambridge Tracts in Mathematics No. 68, London: Cambridge Univ. Press 1974

(1976) Random Fields. Springer Lecture Notes in Mathematics 534

(1979) Canonical and Microcanonical Gibbs States. Z. Wahrscheinlichkeitstheorie u. verw. Gebiete 46, 125-158

Presutti, E., Pulvirenti, M., Tirozzi, B.

(1976) Time Evolution of Infinite Classical Systems with Singular, Long Range, Two Body Interactions. Commun. math. Phys. 47, 81-95

Pulvirenti, M.

(1977) Stability, Equilibrium and KMS for an Infinite Classical System. J. Math. Phys. 18, 2o99-21o3

Pulvirenti, M., Riela, G.

(1977) KMS Condition for Stable States of Infinite Classical Systems. J. Math. Phys. 18, 2364-2367

Reed, M., Simon, B.

(1972) Methods of Modern Mathematical Physics I: Functional Analysis. New York: Academic Press 1972

Ruelle, D.

(1969) Statistical Mechanics. Rigorous Results. New York - Amsterdam: W.A. Benjamin 1969

(197o) Superstable Interaction in Statistical Mechanics. Commun. math. Phys. 18, 127-159

Shiga, T.

(1977) Some Problems Related to Gibbs States, Canonical Gibbs States and Markovian Time Evolutions. Z. Wahrscheinlichkeitstheorie u. verw. Gebiete 39, 339-352

(1979) A Remark on Infinite-Dimensional Wiener Processes with Interactions. Z. Wahrscheinlichkeitstheorie u. verw. Gebiete 47, 299-3o4

Shilov, G.E.

(1968) Generalized Functions and Partial Differential Equations. New York: Gordon and Breach 1968

Simons, G.

(1978) Some Extensions of the Hewitt-Savage Zero-One Law. Z. Wahrscheinlichkeitstheorie u. verw. Gebiete 42, 167-173

Spitzer, F.

(197o) Interaction of Markov Processes. Advances in Math. 5, 246-29o

Sullivan, W.G.

(1973) Potentials for almost Markovian Random Fields. Commun. math. Phys. 33, 61-74

(1975) Markov Processes for Random Fields. Communications of the Dublin Institute for Advanced Studies, series A, No. 23

(1976) Specific Information Gain for Interacting Markov Processes. Z. Wahrscheinlichkeitstheorie u. verw. Gebiete 37, 77-9o

Thompson, R.L.

(1974) Equilibrium States in Thin Energy Shells. Memoirs Amer. Math. Soc. 15o

# Index

Vol. 609: General Topology and Its Relations to Modern Analysis and Algebra IV. Proceedings 1976. Edited by J. Novák. XVIII, 225 pages. 1977.

Vol. 610: G. Jensen, Higher Order Contact of Submanifolds of Homogeneous Spaces. XII, 154 pages. 1977.

Vol. 611: M. Makkai and G. E. Reyes, First Order Categorical Logic. VIII, 301 pages. 1977.

Vol. 612: E. M. Kleinberg, Infinitary Combinatorics and the Axiom of Determinateness. VIII, 150 pages. 1977.

Vol. 613: E. Behrends et al., $L^P$-Structure in Real Banach Spaces. X, 108 pages. 1977.

Vol. 614: H. Yanagihara, Theory of Hopf Algebras Attached to Group Schemes. VIII, 308 pages. 1977.

Vol. 615: Turbulence Seminar, Proceedings 1976/77. Edited by P. Bernard and T. Ratiu. VI, 155 pages. 1977.

Vol. 616: Abelian Group Theory, 2nd New Mexico State University Conference, 1976. Proceedings. Edited by D. Arnold, R. Hunter and E. Walker. X, 423 pages. 1977.

Vol. 617: K. J. Devlin, The Axiom of Constructibility: A Guide for the Mathematician. VIII, 96 pages. 1977.

Vol. 618: I. I. Hirschman, Jr. and D. E. Hughes, Extreme Eigen Values of Toeplitz Operators. VI, 145 pages. 1977.

Vol. 619: Set Theory and Hierarchy Theory V, Bierutowice 1976. Edited by A. Lachlan, M. Srebrny, and A. Zarach. VIII, 358 pages. 1977.

Vol. 620: H. Popp, Moduli Theory and Classification Theory of Algebraic Varieties. VIII, 189 pages. 1977.

Vol. 621: Kauffman et al., The Deficiency Index Problem. VI, 112 pages. 1977.

Vol. 622: Combinatorial Mathematics V, Melbourne 1976. Proceedings. Edited by C. Little. VIII, 213 pages. 1977.

Vol. 623: I. Erdelyi and R. Lange, Spectral Decompositions on Banach Spaces. VIII, 122 pages. 1977.

Vol. 624: Y. Guivarc'h et al., Marches Aléatoires sur les Groupes de Lie. VIII, 292 pages. 1977.

Vol. 625: J. P. Alexander et al., Odd Order Group Actions and Witt Classification of Innerproducts. IV, 202 pages. 1977.

Vol. 626: Number Theory Day, New York 1976. Proceedings. Edited by M. B. Nathanson. VI, 241 pages. 1977.

Vol. 627: Modular Functions of One Variable VI, Bonn 1976. Proceedings. Edited by J.-P. Serre and D. B. Zagier. VI, 339 pages. 1977.

Vol. 628: H. J. Baues, Obstruction Theory on the Homotopy Classification of Maps. XII, 387 pages. 1977.

Vol. 629: W. A. Coppel, Dichotomies in Stability Theory. VI, 98 pages. 1978.

Vol. 630: Numerical Analysis, Proceedings, Biennial Conference, Dundee 1977. Edited by G. A. Watson. XII, 199 pages. 1978.

Vol. 631: Numerical Treatment of Differential Equations. Proceedings 1976. Edited by R. Bulirsch, R. D. Grigorieff, and J. Schröder. X, 219 pages. 1978.

Vol. 632: J.-F. Boutot, Schéma de Picard Local. X, 165 pages. 1978.

Vol. 633: N. R. Coleff and M. E. Herrera, Les Courants Résiduels Associés à une Forme Méromorphe. X, 211 pages. 1978.

Vol. 634: H. Kurke et al., Die Approximationseigenschaft lokaler Ringe. IV, 204 Seiten. 1978.

Vol. 635: T. Y. Lam, Serre's Conjecture. XVI, 227 pages. 1978.

Vol. 636: Journées de Statistique des Processus Stochastiques, Grenoble 1977, Proceedings. Edité par Didier Dacunha-Castelle et Bernard Van Cutsem. VII, 202 pages. 1978.

Vol. 637: W. B. Jurkat, Meromorphe Differentialgleichungen. VII, 194 Seiten. 1978.

Vol. 638: P. Shanahan, The Atiyah-Singer Index Theorem, An Introduction. V, 224 pages. 1978.

Vol. 639: N. Adasch et al., Topological Vector Spaces. V, 125 pages. 1978.

Vol. 640: J. L. Dupont, Curvature and Characteristic Classes. X, 175 pages. 1978.

Vol. 641: Séminaire d'Algèbre Paul Dubreil, Proceedings Paris 1976–1977. Edité par M. P. Malliavin. IV, 367 pages. 1978.

Vol. 642: Theory and Applications of Graphs, Proceedings, Michigan 1976. Edited by Y. Alavi and D. R. Lick. XIV, 635 pages. 1978.

Vol. 643: M. Davis, Multiaxial Actions on Manifolds. VI, 141 pages. 1978.

Vol. 644: Vector Space Measures and Applications I, Proceedings 1977. Edited by R. M. Aron and S. Dineen. VIII, 451 pages. 1978.

Vol. 645: Vector Space Measures and Applications II, Proceedings 1977. Edited by R. M. Aron and S. Dineen. VIII, 218 pages. 1978.

Vol. 646: O. Tammi, Extremum Problems for Bounded Univalent Functions. VIII, 313 pages. 1978.

Vol. 647: L. J. Ratliff, Jr., Chain Conjectures in Ring Theory. VIII, 133 pages. 1978.

Vol. 648: Nonlinear Partial Differential Equations and Applications, Proceedings, Indiana 1976–1977. Edited by J. M. Chadam. VI, 206 pages. 1978.

Vol. 649: Séminaire de Probabilités XII, Proceedings, Strasbourg, 1976–1977. Edité par C. Dellacherie, P. A. Meyer et M. Weil. VIII, 805 pages. 1978.

Vol. 650: C*-Algebras and Applications to Physics. Proceedings 1977. Edited by H. Araki and R. V. Kadison. V, 192 pages. 1978.

Vol. 651: P. W. Michor, Functors and Categories of Banach Spaces. VI, 99 pages. 1978.

Vol. 652: Differential Topology, Foliations and Gelfand-Fuks-Cohomology, Proceedings 1976. Edited by P. A. Schweitzer. XIV, 252 pages. 1978.

Vol. 653: Locally Interacting Systems and Their Application in Biology. Proceedings, 1976. Edited by R. L. Dobrushin, V. I. Kryukov and A. L. Toom. XI, 202 pages. 1978.

Vol. 654: J. P. Buhler, Icosahedral Golois Representations. III, 143 pages. 1978.

Vol. 655: R. Baeza, Quadratic Forms Over Semilocal Rings. VI, 199 pages. 1978.

Vol. 656: Probability Theory on Vector Spaces. Proceedings, 1977. Edited by A. Weron. VIII, 274 pages. 1978.

Vol. 657: Geometric Applications of Homotopy Theory I, Proceedings 1977. Edited by M. G. Barratt and M. E. Mahowald. VIII, 459 pages. 1978.

Vol. 658: Geometric Applications of Homotopy Theory II, Proceedings 1977. Edited by M. G. Barratt and M. E. Mahowald. VIII, 487 pages. 1978.

Vol. 659: Bruckner, Differentiation of Real Functions. X, 247 pages. 1978.

Vol. 660: Equations aux Dérivée Partielles. Proceedings, 1977. Edité par Pham The Lai. VI, 216 pages. 1978.

Vol. 661: P. T. Johnstone, R. Paré, R. D. Rosebrugh, D. Schumacher, R. J. Wood, and G. C. Wraith, Indexed Categories and Their Applications. VII, 260 pages. 1978.

Vol. 662: Akin, The Metric Theory of Banach Manifolds. XIX, 306 pages. 1978.

Vol. 663: J. F. Berglund, H. D. Junghenn, P. Milnes, Compact Right Topological Semigroups and Generalizations of Almost Periodicity. X, 243 pages. 1978.

Vol. 664: Algebraic and Geometric Topology, Proceedings, 1977. Edited by K. C. Millett. XI, 240 pages. 1978.

Vol. 665: Journées d'Analyse Non Linéaire. Proceedings, 1977. Edité par P. Bénilan et J. Robert. VIII, 256 pages. 1978.

Vol. 666: B. Beauzamy, Espaces d'Interpolation Réels: Topologie et Géometrie. X, 104 pages. 1978.

Vol. 667: J. Gilewicz, Approximants de Padé. XIV, 511 pages. 1978.

Vol. 668: The Structure of Attractors in Dynamical Systems. Proceedings, 1977. Edited by J. C. Martin, N. G. Markley and W. Perrizo. VI, 264 pages. 1978.

Vol. 669: Higher Set Theory. Proceedings, 1977. Edited by G. H. Müller and D. S. Scott. XII, 476 pages. 1978.